D1063169

About Island Press

Island Press is the only nonprofit organization in the United States whose principal purpose is the publication of books on environmental issues and natural resource management. We provide solutions-oriented information to professionals, public officials, business and community leaders, and concerned citizens who are shaping responses to environmental problems.

In 2002, Island Press celebrates its eighteenth anniversary as the leading provider of timely and practical books that take a multidisciplinary approach to critical environmental concerns. Our growing list of titles reflects our commitment to bringing the best of an expanding body of literature to the environmental community throughout North America and the world.

Support for Island Press is provided by The Nathan Cummings Foundation, Geraldine R. Dodge Foundation, Doris Duke Charitable Foundation, Educational Foundation of America, The Charles Engelhard Foundation, The Ford Foundation, The George Gund Foundation, The Vira I. Heinz Endowment, The William and Flora Hewlett Foundation, Henry Luce Foundation, The John D. and Catherine T. MacArthur Foundation, The Andrew W. Mellon Foundation, The Moriah Fund, The Curtis and Edith Munson Foundation, National Fish and Wildlife Foundation, The New-Land Foundation, Oak Foundation, The Overbrook Foundation, The David and Lucile Packard Foundation, The Pew Charitable Trusts, The Rockefeller Foundation, The Winslow Foundation, and other generous donors.

The opinions expressed in this book are those of the author(s) and do not necessarily reflect the views of these foundations.

About SCOPE

The Scientific Committee on Problems of the Environment (SCOPE) was established by the International Council for Science (ICSU) in 1969. It brings together natural and social scientists to identify emerging or potential environmental issues and to address jointly the nature and solution of environmental problems on a global basis. Operating at an interface between the science and decision-making sectors, SCOPE's interdisciplinary and critical focus on available knowledge provides analytical and practical tools to promote further research and more sustainable management of the Earth's resources. SCOPE's members, forty national science academies and research councils and twenty-two international scientific unions, committees, and societies, guide and develop its scientific program.

SCOPE 60

Resilience and the Behavior of Large-Scale Systems

THE SCIENTIFIC COMMITTEE ON PROBLEMS OF THE ENVIRONMENT (SCOPE)

SCOPE Series

SCOPE 1–59 in the series were published by John Wiley & Sons, Ltd., U.K. Island Press is the publisher for SCOPE 60 as well as subsequent titles in the series.

SCOPE 60: *Resilience and the Behavior of Large-Scale Systems,* edited by Lance H. Gunderson and Lowell Pritchard Jr.

SCOPE 60

Resilience and the Behavior of Large-Scale Systems

Edited by
Lance H. Gunderson and Lowell Pritchard Jr.

A project of SCOPE, the Scientific Committee on
Problems of the Environment, of the
International Council for Science

ISLAND PRESS
Washington • Covelo • London

Copyright © 2002 Scientific Committee on Problems of the Environment (SCOPE)

All rights reserved under International and Pan-American Copyright Conventions. No part of this book may be reproduced in any form or by any means without permission in writing from the publisher: Island Press, 1718 Connecticut Avenue, N.W., Suite 300, Washington, DC 20009.

Permission requests to reproduce portions of this book should be addressed to SCOPE (Scientific Committee on Problems of the Environment, 51 Boulevard de Montmorency, 75016 Paris, France).

Inquiries regarding licensing publication rights to this book as a whole should be addressed to Island Press (1718 Connecticut Avenue, N.W., Suite 300, Washington, DC 20009, USA).

ISLAND PRESS is a trademark of The Center for Resource Economics.

Library of Congress Cataloging-in-Publication Data

Resilience and the behavior of large-scale systems / Lance H. Gunderson and Lowell Pritchard Jr., editors.
    p. cm. — (SCOPE 60)
"A project of Scientific Committee on Problems of the Environment, International Council of Scientific Unions."
Includes bibliographical references and index
  ISBN 1-55963-970-9 (cloth : alk. paper) — ISBN 1-55963-971-7 (pbk. : alk. paper)
  1. Ecology. 2. Nature—Effect of human beings on. 3. Sustainable development. 4. Social ecology. I. Gunderson, Lance H. II. Pritchard, Lowell. III. International Council of Scientific Unions. Scientific Committee on Problems of the Environment. IV. SCOPE report 60.
  QH541 .R45 2002
    577—dc21   2002010630

British Cataloguing-in-Publication Data available.

Printed on recycled, acid-free paper ♻

Manufactured in the United States of America
09 08 07 06 05 04 03 02   8 7 6 5 4 3 2 1

Contents

Part I.   Understanding Resilience: Theory, Metaphors, and Frameworks

Part III. Summary

List of Figures and Tables

Figures

Tables

Foreword

The Scientific Committee on Problems of the Environment (SCOPE) is one of twenty-six interdisciplinary bodies established by the International Council for Science (ICSU) to address cross-disciplinary issues. SCOPE was established by ICSU in 1969 in response to environmental concerns emerging at that time, in recognition that many of these concerns required scientific input spanning several disciplines represented within its membership. Representatives of forty member countries and twenty-two international, disciplinary-specific unions, scientific committees, and associates currently participate in the work of SCOPE, which directs particular attention to developing countries. The mandate of SCOPE is to assemble, review, and synthesize the information available on environmental changes attributable to human activity and the effects of these changes on humans; to assess and evaluate the methodologies of measurement of environmental parameters; to provide an intelligence service on current research; and to provide informed advice to agencies engaged in studies of the environment.

This synthesis volume discusses the practical consequences of resilience and its role in sustainability of ecosystems affected by humans. Resilience is a measure of the sensitivity of the self-organized spatial and temporal patterns of ecosystems to changes in the intensity or variability of different types and scales of stress. The practical dimensions of resilience are embedded in discussions of how people interact to either enhance or erode resilience.

JOHN W. B. STEWART, Editor-in-Chief

SCOPE Secretariat
51 Boulevard de Montmorency, 75016 Paris, France
Executive Director: Véronique Plocq Fichelet

Preface

The idea for this project arose from a meeting on the island of Askö in the Swedish archipelago, sponsored by the Beijer International Institute for Ecological Economics. The participants recognized the potential of the term *resilience* as a unifying concept in both ecological and social systems. They also recognized that the concepts associated with ecological resilience (alternative stable states, disturbances) were for the most part largely undemonstrated for both ecological and social systems. The purpose of this volume is to help test, refine, and perhaps redefine these notions of ecological resilience.

This book was the result of a series of workshops, the first held on Little St. Simons Island in Georgia, in the United States, the second at Malilangwe Reserve in southeastern Zimbabwe, and the third in Stockholm, Sweden.

In these workshops, we sought to test propositions that described multiple stable states in a wide range of ecosystem types. We hoped to put our conceptual models of resilience at risk by applying these ideas to terrestrial and aquatic systems. As such, we solicited expert ecologists who were knowledgeable about aquatic systems, including freshwater lakes, large inland seas (such as the Baltic), wetlands, and coral reefs. We posed similar propositions to terrestrial ecologists who were experts in semi-arid rangelands, tropical forests, and southern pine forests.

Acknowledgments

This book originated from a proposal submitted to SCOPE entitled "Resilience and the Behavior of Large Scale Systems," written by C. S. Holling and B.-O. Jansson. The approval of that proposal allowed us to pursue and receive generous grants from the John D. and Catherine T. MacArthur Foundation to the University of Florida and the Beijer International Institute for Ecological Economics. Dan Martin, Caren Grown, and Priya Shyamsundar played critical roles as grantors, and we are indebted to them all for supporting the resulting project, "Resilience in Ecosystems, Economies and Institutions." Dan Martin not only approved a grant to support this synthesis, but also facilitated and advanced it because of his abiding conviction that integrative theory must be the foundation for sustainable practice. Other fiscal support was provided from the University of Florida, Emory University, and Wallenberg Foundation. Specific acknowledgements are also listed within chapters. Beverly Gunderson produced much of the final artwork for this volume, and Pille Bunnell contributed to the graphics in chapters 1 and 10. Versions of chapters 2 and 3 have appeared in the journal *Conservation Ecology*.

The editors are greatly indebted to C. S. (Buzz) Holling and B.-O. Jansson. Bengt-Owe recognized the global significance and the theoretically fundamental value of testing and publishing on these concepts. Most of what is contained here was spawned by Buzz's brilliant insight almost thirty years ago—which he said started with an argument over a possible glitch in a computer program. Out of that argument came his insight into and definition of ecological resilience—just one of his many creative contributions to the theory and practice of ecology in the twentieth century.

PART I
Understanding Resilience: Theory, Metaphors, and Frameworks

1

Resilience of Large-Scale Resource Systems

Lance H. Gunderson, C. S. Holling,
Lowell Pritchard Jr., and Garry D. Peterson

Regional-scale systems of people and nature provide some of the most vexing challenges for attaining social goals of sustainability, biological conservation, or economic development. There are many more examples of failures than successes, as measured by numerous resource systems that exist in a constant or recurring state of crisis (Ludwig et al. 1993). In the Florida Everglades, agricultural interests, environmentalists, and urban residents contest with one another for control over clean water (Light et al. 1995). In the Pacific Northwest region of the United States, various advocates of salmon argue over the appropriate use of the Columbia River with those who prefer cheap hydroelectric power (Lee 1993; Volkman and McConnaha 1993). The nations surrounding the Baltic Sea struggle with issues of governance as the fish populations and water quality of the sea declines (Jansson and Velner 1995). Within Zimbabwe, large-scale land use conversions are testing stabilities of both ecological and political structures. In these cases resource management has taken a pathological form in which the complexity of the issues, institutional inertia, and uncertainty lead to a state of institutional gridlock, when inaction causes ecological issues to be ignored and existing policies and relationships to be continued.

Paradoxically, this failure often arises from the success of initial management actions. Managers of natural resource systems are often successful at rapidly achieving a set of narrowly defined goals. Unfortunately, this success encourages people to build up a dependence upon its continuation while simultaneously eroding away the ecological support that it requires. This leads to a state in which ecological change is increasingly undesirable to the people dependent

3

upon the natural resource and simultaneously more difficult to avoid. This management pathology leads to unwanted changes in nature, a loss of ecological resilience, conservative management policies, and loss of trust in management agencies.

Recent work reveals a way out of this pathology in large, regional-scale systems. These systems move through periods of surprise, crisis, and reformation (Gunderson et al. 1995). Managers are surprised when the inadequacies of many, if not most, management policies are revealed by ecosystem dynamics. A crisis occurs when it is becomes unambiguously clear that existing policies caused this surprise. The crisis is followed by periods of denial, resistance, and often, finally, by a period of reformation during which new policies are developed and implemented. It is during these periods of crisis that institutions and the connections between them are most open to dramatic transformation. This ability to transform and survive requires that the resource system have sufficient resilience to permit the experimental development of new management policies.

What Is Resilience?

Resilience has been defined in two different ways in the ecological literature, each reflecting different aspects of stability. One definition focuses on efficiency and depends on constancy and predictability—all attributes of engineers' desire for fail-safe design. The other focuses on persistence, despite change and unpredictability—all attributes embraced and celebrated by evolutionary biologists and by resource managers who search for safe-fail designs. Holling (1973) first emphasized these contrasting aspects of stability to draw attention to the tensions between efficiency and persistence, between constancy and change, and between predictability and unpredictability.

The more common definition, which we term *engineering resilience* (Holling 1996), conceives ecological systems to exist close to a stable steady state. Engineering resilience, then, is the speed of return to the steady state following a perturbation (Pimm 1984; O'Neill et al. 1986; Tilman and Downing 1994). This idea of disturbance away from and return to a stable state is also at the center of twentieth-century economic theory (Varian 1992; Kamien and Schwartz 1991).

The second definition, which we term *ecological resilience* (Walker et al. 1981; Holling 1996), emphasizes conditions far from any stable steady state, where instabilities can shift or flip a system into another regime of behavior—in other words, to another stability domain (Holling 1973). In this case, resilience is measured by the magnitude of disturbance that can be absorbed before the system is restructured with different controlling variables and processes.

The differences between these two aspects of stability—essentially between a focus on maintaining efficiency of function (engineering resilience) and a focus on maintaining existence of function (ecological resilience)—are so fundamental that they can become alternative paradigms in which subscribers dwell on received wisdom rather than the reality of nature. Those using the concept of engineering resilience tend to explore system behavior near a known stable state, while those examining ecological resilience tend to search for alternative stable states and the properties of the boundaries between states.

Those who explore engineering resilience and the near-equilibrium behavior of ecosystems operate in the primarily deductive tradition of mathematical theory (e.g., Pimm 1984) that imagines simplified, untouched ecological systems; or they draw upon the traditions of engineering, which are motivated by the need to design systems with a single operating objective (Waide and Webster 1976; DeAngelis 1980; O'Neill et al. 1986). These approaches simplify the mathematics and accommodate the engineer's drive to develop optimal designs. However, there is an implicit assumption that ecosystems exhibit only one equilibrium steady state or, if other operating states exist, that those states should be avoided (figure 1.1).

On the other hand, those who emphasize ecological resilience come from

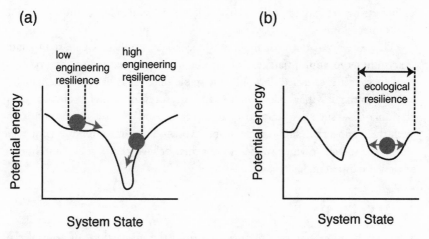

Figure 1.1. Stability landscapes can be used to represent the dynamics of a system and alternative definitions of resilience. The ball represents the system "state." The state can be changed by disturbances, which move the system along a stability landscape. The shape of the landscape is determined by controlling variables of the system. Engineering resilience (speed of recovery) is a local measure and is determined by the slope of the landscape. *(a)* Depressions in the landscape with low slopes have less engineering resilience than areas that have steep slopes. *(b)* Ecological resilience of a system corresponds to the width of a stability basin.

traditions of applied mathematics and applied resource ecology at the scale of ecosystems—for example, of the dynamics and management of freshwater systems (Fiering 1982), of forests (Holling et al. 1977), of fisheries (Walters 1986), of semi-arid grasslands (Walker et al. 1981) and of interacting populations in nature (Sinclair et al. 1990; Dublin et al. 1990). Because these researchers are rooted in inductive rather than deductive theory formation, and because they have experience with the impacts of large-scale management actions, they believe that it is the variability of critical variables that forms and maintains the stability landscape. When this variability is reduced, an ecosystem can flip from one organization to another (figure 1.1).

In economics, there has also been a focus on single stable state. The history of economics has been to rapidly move from establishing the existence of a general equilibrium to examining issues of equilibrium uniqueness, stability, and comparative statics. If multiple equilibria are shown to theoretically exist, then the challenge is to theoretically reduce the salience of alternate stable states by proposing that expectations, norms, and social institutions make some equilibria unlikely. This approach does not examine or explain the conditions that can cause a system to move from one stability domain to another. Recently, however, the identification of multi-stable states due to path dependence (Arthur et al. 1987), chreodic development (Clark and Juma 1987), and non-convexities such as increasing returns to scale (David 1985) has reintroduced multiple stable states to economics.

The existence, or at least the importance, of multiple or single stable states determines the appropriateness of an engineering or ecological approach to resilience. If it is assumed that only one stable state exists or can be designed to exist, then the only possible definition and measures for resilience are near-equilibrium ones—such as characteristic return time. And that is certainly consistent with the engineer's desire to make things work—and not to intentionally make things that break down or suddenly shift their behavior. But nature and human society are different.

Why Study Resilience?

Complex resource systems are organized from the interactions of a set of ecological, social, and economic systems across a range of scales. Resilience is central to understanding the dynamics of these systems and their vulnerability to various shocks and disruptions. Resilience measures the strength of mutual reinforcement between processes, incorporating both the ability of a system to persist despite disruptions and the ability to regenerate and maintain existing organization. Resilience allows a system to withstand the failure of management actions. Management is necessarily based upon incomplete understanding, and therefore ecological resilience allows people in resource systems the opportunity to learn and change.

The importance of the role of resilience in ecosystems, flexibility of institutions, and incentives in economies emerged in a sequence of meetings held on the island of Askö in the Swedish archipelago. Sponsored by the Beijer International Institute for Ecological Economics, these meetings brought together economists and natural scientists to explore similarities and differences in views and experiences of change. Their conclusions were that economic growth is not inherently good, nor inherently bad, but that economic growth cannot in the long term compensate for declines in environmental quality. They also concluded that the growing scale of human activities is encountering the limits of nature to sustain that expansion (Folke and Berkes 1998; Arrow et al. 1995).

The familiar responses to these issues are often flawed, because the theories of change underlying them are inadequate. The stereotypical economist might say "get the prices right" (i.e., ensure that prices internalize significant environmental externalities) without recognizing that price systems require a stable context where social and ecosystem processes behave "nicely" in a mathematical sense (i.e., are continuous and convex). The stereotypical social scientist might say "get the institutions right" without comprehending the degree to which those institutions submerge ecological uncertainties and economic and political interests. The stereotypical ecologist might say "get the indicators right" without recognizing the surprises that nature and people inexorably and continuously generate. And the stereotypical engineer might say "get the technological control right and we can eliminate those surprises" without recognizing the limits to knowledge and control imposed by the inherent uncertainty and unpredictability of the ever-evolving interaction of people and nature.

Although based on bad or insufficient theory, such simple prescriptions are attractive because they seem to replace inherent uncertainty with the spurious certitude of ideology, of precise numbers or of action. The theories implicit in these examples ignore multi-stable states. They ignore the possibility that the slow erosion of key controlling processes can cause an ecosystem or economy to abruptly flip into a different state that might effectively be irreversible. In an ecosystem, this might be caused by the gradual loss of a species in a keystone set that together determine structure and behavior over specific ranges of scale. In a resource-based economy, it might be implementation of maximum sustained yield policies that reduce spatial diversity, evolve ever-narrower economic dependencies, and develop more rigid organizations. In an economy, it might be caused by the channeling of loans through personal networks, allowing bad loans to accumulate to such a point that they cause an entire banking and finance system to collapse—such as the Asian financial crisis in the late 1990s.

It increasingly appears that effective and sustainable development of technology, institutions, economies, and ecosystems requires ways to deal not only with near equilibrium efficiency but also with the reality of more than one possible equilibrium. If there are multiple equilibria, in which direction should the finger on the invisible hand of Adam Smith point? If there is more than one objective function, where does the engineer search for optimal designs? In such a context, a near-equilibrium approach is myopic. Attention should shift to determining the constructive role of instability in maintaining diversity and persistence and to management designs that maintain ecosystem function despite unexpected disturbances. Such designs maintain or expand the ecological resilience of those ecological "services" that invisibly provide the foundations for sustaining economic activity and human society.

The goal of this volume is to begin to understand how the properties of ecological resilience and human adaptability interact in complex, large systems (regional scale). To lay a foundation for this volume, we initially review other key properties of complex adaptive systems that contribute to resilience.

Properties of Complex Adaptive Systems

We propose that the behavior of complex adaptive systems depends upon four key properties: ecological resilience, complexity, self-organization, and order. As discussed above, resilience is the extent to which a system can withstand disruption before shifting into another state. Complexity is the variety of structures and processes that occur within a system. Self-organization is the ability of these structures and processes to mutually interact to reinforce and sustain each other. The process of self-organization produces order from disorder, but the interaction of processes across scales also destroys, and reconfigures, ecological organization, producing complex ecological dynamics. The next three sections elaborate upon the role these properties play in complex systems, and how these other properties contribute and interact with resilience.

Diversity and Stability

The relationship between biological diversity and ecological stability has been an ongoing debate in ecology since the time of Darwin (1860; also Elton 1958; May 1973; Tilman and Downing 1994, 1996). The question is whether an ecosystem that includes more species is more stable than one that includes fewer species?

Tilman and Downing (1994) and Tilman (1996) demonstrated that an increase in species number increases the efficiency and stability of some ecosystem functions but decreases the stability of the populations of the species, at least over ecologically brief periods. Although this work is important and interesting, it focuses only on the behavior of ecosystems near some steady state. But, as we've discussed above, we feel it is important to discover the role of ecological diversity over a much broader range of variations. This is where the relationship between diversity and resilience has been poorly developed.

When grappling with this broader relationship between diversity and resilience, most turn to two commonly discussed hypotheses: Ehrlich's (1991) *rivet hypothesis* and Walker's (1992) *driver and passengers hypothesis.* The rivet hypothesis proposes that there is little change in ecosystem function as species are added or lost, until a threshold is reached. At that threshold the addition or removal of a single species leads to system reorganization (just as popping rivets from a seam causes little change at first, but at some point sudden, disastrous change will occur). The rivet hypothesis assumes that species have overlapping roles and that as species are lost the ecological resilience of the system is decreased, and then overcome entirely. Walker proposes that species can be divided into *functional groups,* or *guilds,* which are groups of species that act in an ecologically similar way. Walker proposes that these groups can be divided into "drivers" and "passengers." Drivers are "keystone" species that control the future of an ecosystem, while the passengers live in but do not significantly alter their ecosystem. However, as conditions change, endogenously or exogenously, species shift roles. Removing passengers has little effect, while removing drivers can have a large impact. Ecological resilience resides both in the diversity of the drivers and in the number of passengers who are potential drivers. These two hypotheses provide a start, but richer models of ecological complexity are needed that better incorporate ecological processes, dynamics, and scale.

Ecosystems are resilient when ecological interactions reinforce one another and dampen disruptions. Such situations may arise due to compensation when a species with an ecological function similar to another species increases in abundance as the other declines (Holling 1996) or as one species reduces the impact of a disruption on other species.

Theory, models, and data suggest that a small number of keystone processes create discontinuous spatial and temporal patterns in ecosystems (Holling et al. 1996; Levin 1995) yet allow for great diversity of organisms. Such keystone ecological processes produce a discontinuous distribution of structures in ecosystems, and these discontinuous structures generate discontinuous patterns in adult body masses of animals that inhabit landscapes (Holling 1992; Morton 1990; Allen et al. 1999). Consequently, while animals that function at the same scale are separated by functional specialization (e.g., insectivores, herbivores, arboreal frugivores, etc.), animals that function at different scales can utilize similar resources (e.g., shrews and anteaters are both insectivores but utilize insects at different scales). We propose that the resilience of ecological processes, and therefore of the ecosystems they maintain, depends upon the distribution of functional groups within and across scales (Peterson et al. 1998).

Across-scale resilience is produced by the replication of process at different scales. The apparent redundancy of similar functions replicated at different scales adds resilience to an ecosystem. Because most disturbances occur at specific scales, similar functions that operate at other scales are maintained.

Local processes such as competitive relationships certainly contribute to species differences among ecosystems. However, the structural differences among ecosystems from the tundra to the tropics are primarily produced by larger-scale disturbance processes that are initiated locally and then spread across landscapes. These contagious processes include abiotic processes, such as fire, storms, and floods, and zootic processes, such as insect outbreaks, large mammal herbivory, and habitat modification (Naiman 1988; McNaughton 1988; Pastor and Cohen 1996). These processes, interacting with topography and regional climate, form the ecosystem-specific structures that shape the morphology and diversity of animal communities. They also generate spatial and temporal variation that increases the diversity of plant species by periodically overriding the competitive dominance relations that occur locally (Holling 1991). For example, in the eastern boreal forest of Canada, fire and spruce-budworm outbreaks kill large areas of forest. Through interactions with climate, existing vegetation, and each other, these processes produce a mosaic of even-aged forest stands in the landscape. Since the age a stand reaches before being destroyed is primarily determined by disturbance, and what species exist within the stand is influenced by landscape pattern, these disturbance processes also strongly control what exists within stands. Consequently, these disturbance processes strongly influence the distribution and type of resources that occur in eastern Canadian boreal forest across a broad range of ecological scales.

An ecosystem that has several scales of ecological structure allows members of multi-taxa food guilds to minimize competition by utilizing resources that are available at different scales (figure 1.2). The replication of function across scales can be seen on Brazil's Maracá Island Ecological Reserve, where palm seeds are dispersed across a range of scales by a variety of species (Fragoso 1997). Seed dispensers range in size from small rodents, which typically disperse seeds within 5 meters of parent trees, to tapirs (*Tayassu tajacu*), which disperse seeds as far away as 2 kilometers. Seed dispersal at multiple scales allows the palm population to

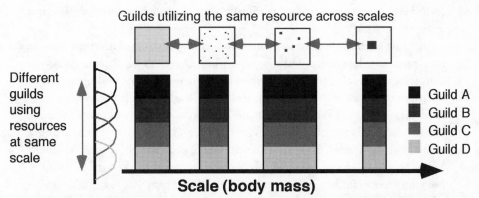

Figure 1.2. Animal species belonging to different ecological guilds exist at different body sizes. For example, there are both small and large insectivores. This distribution provides two forms of resilience. At the same scale animals from different guilds can utilize the same resources with lower efficiency. Also, animals that utilize the same resources can begin to utilize resources from a lower level if they form large enough aggregations. For example, if insectivores were removed from a group, insects would become easier to catch, making it worthwhile for animals at the same scale to switch from their normal food to insects, and it may become worthwhile for larger insectivores to eat prey items they normally would not eat.

persist despite a variety of disturbance processes occurring at different scales, because the trees are dispersed across the landscape at different scales.

Within-scale resilience complements cross-scale resilience. Within-scale resilience is produced by compensating overlap of ecological function between similar processes that occur at the same scales. For example, when a range of food resources is exploited by a set of foragers, rapid response to sudden increases or decreases in one type of food becomes possible and introduces strong negative feedback regulation over a wide range of densities of the food items (Holling 1987). The consequence of all that variety is that the species combine to form an overlapping set of reinforcing influences that are less like the redundancy of engineered devices and more like portfolio diversity strategies of investors. The risks and benefits are spread widely to retain overall consistency in performance independent of wide fluctuations in the individual species. Functional diversity provides great robustness to the functioning of the process and, as a consequence, provides great resilience to the system behavior. Moreover, this seems to be the way many biological processes are regulated: overlapping influences by multiple processes each one of which is inefficient in its individual effect but together operating in a robust manner. For example, such multiple-mechanism features control body temperature regulation in endotherms, depth perception in animals with binocular vision, and direction in bird migration.

Because of the ecological resilience produced by functional diversity and the nonlinear way behavior suddenly flips from one ecological organization to another, gradual loss of species involved in maintaining ecological organization initially may have little immediate impact. But, as the loss of species continues, different behavior will emerge more and more frequently in an increasing number of places. To the observer, it would appear as if only the few remaining species were critical when in fact all contribute to ecosystem resilience. Although behavior would change suddenly, resilience measured as the size of stability domains (*sensu* Holling 1973) would gradually contract. The system, in gradually losing resilience, would become increasingly vulnerable to perturbations that earlier could be absorbed without change in function, pattern, and control.

Cross-Scale Dynamics

In nature, different structures and processes dominate at different scales. For example, in the boreal forest, fresh needles cycle yearly, the crown of foliage cycles with a decadal period, and trees, gaps, and stands all cycle at periods close to a century in length and even longer. Ecological organization can be viewed as a hierarchy in which each hierarchical level has its own distinct spatial and temporal attributes. A critical feature of such hierarchies is the asymmetric interactions that occur between levels (Allen and Starr 1982; O'Neill et al. 1986). In particular, the larger, slower levels constrain the behavior of faster levels; that is, slower levels control faster ones. However, if that were the only asymmetry, then hierarchies would be static structures and it would be impossible for organisms to exert control over slower environmental variables. In fact, these hierarchies are not static but are transitory structures maintained by interaction across scales.

Birth, growth, death, and renewal cycles (figure 1.3) transform hierarchies from fixed static structures to dynamic adaptive entities whose levels are sensitive to small disturbances at the transition from growth to collapse (the omega phase) and at the transition from reorganization to rapid growth (the alpha phase). During other times, the processes are stable and resilient. They constrain lower levels and are immune to the buzz of noise from small and faster processes. It is at the two phase transitions between gradual and rapid change that the large and slow entities become sensitive to change from the small and fast ones.

When the system is reaching the limits to its conservative growth, it becomes increasingly brittle and its accumulated capital is ready to fuel rapid structural changes. The system is very stable, but that stability derives from a web of interacting connections. When this tightly connected system is disrupted, the dis-

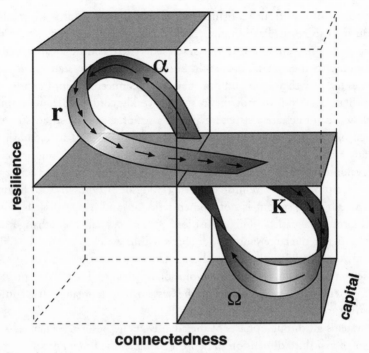

connectedness

Figure 1.3. Ecosystem dynamics, indicating transitions among stages (r, K, Ω, α) and the resilience of each stage. The arrows show the speed of that flow in the cycle, arrows close to each other indicate a rapidly changing situation and arrows far from each other indicate a slowly changing situation. The cycle reflects changes in three attributes: x-axis, the degree of connectedness among variables, y-axis, the resilience of the system (low to high), and z-axis, the amount of accumulated capital (nutrients, carbon) stored in variables that are the dominant structuring variables at that moment in the system.

ruption can spread quickly, destabilizing the entire system. The specific nature and timing of the collapse-initiating disturbance determines, within some bounds, the future trajectory of the system. Therefore, this brittle state presents the opportunity for a change at a small scale to cascade rapidly through a system and bring about its rapid transformation. This is the "revolt of the slave variable" (Diener and Poston 1984). Such a collapse can be initiated by either internal conditions (e.g., the amplification of internal oscillations) or external events (e.g., the amplification of an external disturbance). Internally induced brittleness (linked to overconnected and accumulated capital) provides the conditions for an externally triggered collapse.

The second opportunity for small-scale processes to cause system change is during the transition from reorganization to exploitation, that is, from alpha to r. During this reorganization phase, the system is in a state opposite to that of the conservation phase previously described. There is little local regulation and stability, so the system can easily be moved from one state to another. Resources for growth are present, but they are disconnected from the processes that facilitate and control growth. In such a weakly connected state, a small-scale change can nucleate a structure amidst a sea of chaos. This new structure can then use the available resources to grow explosively and to establish the exploitative path along which the system develops. As in Waddington's chreodic development model, there is not a stable point; rather, there is a stable trajectory that progressively reinforces itself (Hodgson 1993). In Waddington's (1969) words, "the system is not homeostatic (around a point), it is homeorhetic (around a path)." This transition occurs as small-scale changes sow seeds of order in the larger and slower chaos within which they are embedded. The budworm example illustrates these changes where transient bottom-up asymmetry provides an opening for evolutionary change. That is, the previous system pattern may reassert itself, or the system may reorganize itself into a novel structure.

As systems go through phases of the adaptive cycle, resilience changes. This is presented as a third dimension in figure 1.3. When the system is reaching the limits to its conservative growth (K phase), it becomes increasingly brittle and its accumulated capital is ready to fuel rapid structural changes. The system is very stable, but that stability is self-maintaining and brittle, leaving the system vulnerable to novelty. A small disturbance can push it out of that stable domain into catastrophe; hence, its resilience is relatively low. The nature and timing of the collapse-initiating disturbance determines, within limits defined by the nature of lower and higher hierarchical levels, the future trajectory of the system. During reorganization (alpha phase) a system has greater resilience but little stability. Fluctuations in large-scale processes, such as climate, or in small-scale processes, such as a seed bank, can result in a system establishing different organizations.

Panarchy

The accumulating body of evidence from studies of ecosystems indicates that processes and structures often are discontinuous, and ecosystems can exhibit multiple stable states. The four-phase cycle of adaptive renewal captures many of these dynamics for ecological systems (figure 1.3). However, that model is appropriate for structures within a specific range of scales, and those dynamics occur at multiple hierarchical scales or levels. By considering the dynamics of adaptive cycles interacting across scale, we have developed a model of cross-scale ecological organization that we call *panarchy*.

Panarchy describes the dynamic nature of interacting hierarchies. We prefer this term over *hierarchy*, because it emphasizes the dynamic and transient nature of connections between scales. Different systems exhibit the birth, growth, death, and renewal phases of the adaptive cycle at different specific scale ranges (figure 1.3). As in a conventional hierarchy, the processes within a level are stable and self-maintaining, constraining smaller and faster processes while benign and immune to their fast fluctuations. However, the panarchy model transforms hierarchies from fixed static structures to dynamic adaptive entities whose levels are sensitive to small disturbances at the transition from growth to collapse (the omega phase) and during the transition from reorganization to rapid growth (the alpha phase). It emphasizes that the "creative destruction" that follows "revolt" and the "remembrance" that shapes reorganization are the products of cross-scale interactions (figure 1.4).

Tests of resilience are from the "revolt" event. Revolts are enabled by the interaction across scales of slow and fast variables. Resilience shrinks as slow

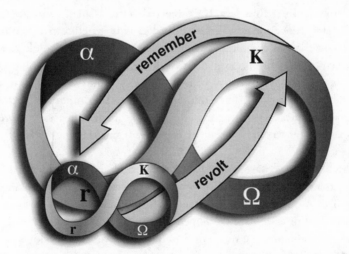

Figure 1.4. Sample of panarchical interactions showing two structural elements (such as a pine needle, represented by the smaller cycle, and tree crown, represented by the larger cycle). For each element, the four system phases (r, K, Ω, α) are as described in figure 1.3. Arrows labeled "revolt" and "remember" indicate key linkages across space and time scales. Smaller-scale elements that are in the Ω phase (creative destruction) can synchronize and cascade to create a transition to the Ω phase at broader scales, as represented by the "revolt" arrow. Broader scales provide resources during smaller-scale reorganization phase, as suggested by the "remember" arrow.

accumulation of capital over time and an increase in connectivity over space makes the system vulnerable to the destabilizing effect of fast variables that trigger a cascade of creative destruction. This has been described as the "revolt of the slave variable." Collapse can be initiated by either internal conditions or external events, but typically it is internally induced brittleness (linked to overconnected and accumulated capital) that sets the conditions for collapse, which coincides with a proximal triggering event associated with larger-scale variation (i.e., external events).

Resilience is reestablished by the "remember" process that connects the present to the past (e.g., accumulated seed banks) and the local to the distant (in-dispersal of propagules). It involves processes of regeneration and renewal. In social systems, examples of these connections include drawing upon and utilizing social capital, traditional knowledge, and wisdom. The remember process is also an opportunity for small-scale processes to cause system change during the transition from reorganization to exploitation, that is, from alpha to r. During this reorganization phase, the system is in a state opposite to that of the conservation phase. There is little local regulation and stability, so the system can easily be moved from one state to another. Resources for growth are present, but they are disconnected from the processes that facilitate and control growth. In such a weakly connected state, a small-scale change can nucleate a structure amidst the sea of disorder. This structure can then use the available resources to grow explosively and to establish the exploitative path along which the system develops and then locks into. This transition occurs as small-scale changes sow seeds of order in the larger and slower chaos within which they are embedded. The transient but critically important bottom-up asymmetry provides an opening for evolutionary change. That is, the previous system pattern may reassert itself, or the system may reorganize itself into a novel structure.

In this section, we have reviewed the theoretical and empirical foundations for understanding resilience in complex, adaptive systems. That foundation will be examined in the remaining chapters of this volume. To provide some focus for such evaluations, we have developed a set of propositions. To provide some structure for evaluating the theoretical foundation, we have developed a set of propositions that describe the main claims of these new theories.

Propositions

A theoretical review of the ecological processes organizing large-scale ecosystems leads to a number of propositions. Because these propositions are largely based upon ecological literature, they may be inappropriate in other disciplinary areas.

For this reason, we offer them provisionally, so that their examination and testing focus the following discussion. Our intent is to provide a framework for subsequent chapters to explore, refine, and reject these constructs as they apply to a wide variety of large-scale, ecological systems. In this spirit, we pose the following propositions:

- *The organization of regional resource systems emerges from the interaction of a few variables.* The essential structure and dynamics of complex systems are produced by the interaction of at least three, but no more than six, variables that operate at spatial and temporal scales that differ by approximately an order of magnitude.
- *Complex systems have multiple stable states.* Complex systems can exhibit alternative stable organizations. Transitions between different organizations are due to changes in the interaction of structuring variables. Change often occurs when gradual change in a slow variable alters the interactions among faster variables.
- *Resilience derives from functional reinforcement across scales and functional overlap within scales.* Resilience derives from both a duplication of function across a range of spatial and temporal scales and a diversity of different functions operating within each scale.
- *Vulnerability increases as sources of novelty are eliminated and as functional diversity and cross-scale functional replication are reduced.* Diminished sources of novelty reduce the ability of a system to recover from disturbances. The elimination of structuring species or processes can cause an ecosystem to reorganize. A reduction in functional diversity and duplication of functions reduces the ability of a system to persist.

Evaluation of Propositions in Large-Scale Ecosystems

These propositions are evaluated in the case studies presented in this book. The book itself is divided into three parts: a theoretical introduction, case studies, and synthesis. The theoretical introduction is provided by this chapter and the next, which use a set of mathematical metaphors to describe and deepen our understanding of the concepts of resilience. The second, and largest, part of this volume is a series of case studies that explore the biophysical dimensions of resilience and evaluate the propositions described above. These case studies review resilience in both terrestrial and aquatic systems. The terrestrial systems include boreal forest, tropical rainforest, tropical dry forest, semi-arid savanna, and tropical agroecosystems. Aquatic ecosystems considered include coral reefs, freshwater lakes, wetlands, and inland seas (specifically the Baltic Sea). The book concludes with a synthesis section, revisiting these propositions in light of the case studies.

Literature Cited

Allen, C. R., E. Forys, and C. S. Holling. 1999. Body mass patterns predict invasions and extinctions in transforming landscapes. *Ecosystems* 2:114–121.

Allen, T. F. H., and T. B. Starr. 1982. *Hierarchy: Perspectives for ecological complexity.* Chicago: University of Chicago Press.

Arthur, W. B., Y. M. Ermoliev, and Y. M. Kaniovski. 1987. Path-dependent processes and the emergence of macro-structure. *European Journal of Operations Research* 30:294–303.

Clark, N., and C. Juma. 1987. *Long-run economics: An evolutionary approach to economics growth.* London: Pinter.

Darwin, C. 1860. *On the origin of species by means of natural selection, or, the preservation of favoured races in the struggle for life.* 5th ed. London: J. Murray.

David, P. A. 1985. Clio and the economics of QWERTY. *American Economic Review* 75:332–337.

DeAngelis, D. L. 1980. Energy flow, nutrient cycling and ecosystem resilience. *Ecology* 61:764–771.

Diener, M., and T. Poston. 1984. On the perfect delay convention or the revolt of the slaved variables. Pp. 249–268 in *Chaos and order in nature*, edited by H. Haken. Berlin: Springer-Verlag.

Dublin, H. T., A. R. E. Sinclair, and J. McGlade. 1990. Elephants and fire as causes of multiple stable states in the Serengeti-Mara woodlands. *Journal of Animal Ecology* 59:1147–1164.

Ehrlich, P. R. 1991. Population diversity and the future of ecosystems. *Science* 254:175.

Elton, C. S. 1958. *The ecology of invasions by animals and plants.* London: Methuen.

Fiering, M. B. 1982. Alternative indices of resilience. *Water Resources Research* 18:33–39.

Folke, C., and F. Berkes, eds. 1998. *Linking ecological and social systems.* Cambridge: Cambridge University Press.

Fragoso, J. M. V. 1997. Tapir-generated seed shadows: Scale-dependent patchiness in the Amazon rain forest. *Journal of Ecology* 85:519–529.

Gunderson, L. H., C. S. Holling, and S. S. Light. *Barriers and bridges to renewal of ecosystems and institutions.* New York: Columbia University Press.

Hodgson, G. M. 1993. *Economics and evolution: Bringing life back into economics.* Ann Arbor: University of Michigan Press.

Holling, C. S. 1973. Resilience and stability of ecological systems. *Annual Review of Ecology and Systematics* 4:1–23.

———. 1991. The role of forest insects in structuring the boreal landscape. Pp. 170–191 in *A systems analysis of the global boreal forest*, edited by H. H. Shugart, R. Leemans, and G. B. Bonan. Cambridge: Cambridge University Press.

———. 1992. Cross-scale morphology, geometry, and dynamics of ecosystems. *Ecological Monographs* 62:447–502.

———. 1996. Engineering resilience versus ecological resilience. Pp. 31–43 in *Engineering within ecological constraints*, edited by P. C. Schulze. Washington, D.C.: National Academy Press.

Holling, C. S., D. D. Jones, and W. C. Clark. 1977. Ecological policy design: A case study of forest and pest management. In *Proceedings of a conference on pest management,* edited by G. A. Norton and C. S. Holling. Oct. 1976, IIASA CP-77-6, 13-90, Laxenburg, Austria.

Holling, C. S., G. Peterson, P. Marples, J. Sendzimir, K. Redford, L. Gunderson, and D. Lambert. 1996. Self-organization in ecosystems: Lumpy geometries, periodicities and morphologies. Pp. 346–384 in *Global change in terrestrial ecosystems,* edited by B. H. Walker and W. L. Steffen. Cambridge: Cambridge University Press.

Jansson, B.-O., and H. Velner. 1995. The Baltic: The sea of surprises. Pp. 292–374 in *Barriers and bridges to renewal of ecosystems and institutions,* edited by L. H. Gunderson, C. S. Holling, and S. S. Light. New York: Columbia University Press.

Kamien, M. I., and N. L. Schwartz. 1991. *Dynamic optimization: The calculus of variations and optimal control in economics and management.* Amsterdam: North-Holland.

Lee, K. N. 1993. *Compass and gyroscope.* Washington, D.C.: Island Press.

Levin, S. 1995. *Biodiversity: Interfacing populations and ecosystems.* Kyoto: Kyoto University Press.

Light, S. S., L. H. Gunderson, and C. S. Holling. 1995. The Everglades: Evolution of management in a turbulent ecosystem. Pp. 103–168 in *Barriers and bridges to renewal of ecosystems and institutions,* edited by L. H. Gunderson, C. S. Holling, and S. S. Light. New York: Columbia University Press.

Ludwig, D., R. Hilborn, and C. Walters. 1993. Uncertainty, resource exploitation, and conservation: Lessons from history. *Science* 260:17, 36.

May, R. M. 1973. *Stability and complexity in model ecosystems.* Princeton, N.J.: Princeton University Press.

McNaughton, S. J., R. W. Ruess, and S. W. Seagle. 1988. Large mammals and process dynamics in African ecosystems. *BioScience* 38:794–800.

Morris, R. F. 1963. The dynamics of epidemic spruce budworm populations. *Memoirs of the Entomological Society of Canada* 21:332.

Morton, S. R. 1990. The impact of European settlement on the vertebrate animals of arid Australia: A conceptual model. *Proceedings of the Ecological Society of Australia* 16:201–213.

Naiman, R. J. 1988. Animal influences on ecosystem dynamics. *Bioscience* 38:750–752.

O'Neill, R. V., D. L. DeAngelis, J. B. Waide, and T. F. H. Allen. 1986. *A hierarchical concept of ecosystems.* Princeton, N.J.: Princeton University Press.

Peterson, G. D., C. R. Allen, and C. S. Holling. 1998. Ecological resilience, biodiversity, and scale. *Ecosystems* 1:6–18.

Pastor, J., and Y. Cohen. 1997. Herbivores, the functional diversity of plant species, and the cycling of nutrients in ecosystems. *Theoretical Population Biology* 51:165–179.

Pimm, S. L. 1984. The complexity and stability of ecosystems. *Nature* 307:321–326.

Sinclair, A. R. E., P. D. Olsen, and T. D. Redhead. 1990. Can predators regulate small mammal populations? Evidence from house mouse outbreaks in Australia. *Oikos* 59:382–392.

Tilman, D. 1996. Biodiversity: Population versus ecosystem stability. *Ecology* 77: 350–363.

Tilman, D., and J. A. Downing. 1994. Biodiversity and stability in grasslands. *Nature* 367:363–365.

Varian, H. R. 1992. *Microeconomic analysis.* New York and London: W. W. Norton.

Volkman, J., and W. E. McConnaha. 1993. Through a glass darkly: Columbia River salmon, the Endangered Species Act, and adaptive management. *Environmental Law* 23:1249–1272.

Waddington, C. H. 1969. The theory of evolution today. Pp. 357–374 in *Beyond reductionism: New perspectives in the life sciences*, edited by A. Koestler and J. R. Smythies. London: Hutchinson.

Waide, J. B., and J. R. Webster. 1976. Engineering systems analysis: Applicability to ecosystems. Pp. 329–371 in *Systems analysis and simulation in ecology*, vol. 4, edited by B. C. Patten. New York: Academic Press.

Walker, B. H. 1992. Biological diversity and ecological redundancy. *Conservation Biology* 6:18–23.

Walker, B. H., D. Ludwig, C. S. Holling, and R. M. Peterman. 1981. Stability of semi-arid savanna grazing systems. *Journal of Ecology* 69:473–498.

Walters, C. J. 1986. *Adaptive management of renewable resources*. New York: McGraw-Hill.

2

Models and Metaphors of Sustainability, Stability, and Resilience

Donald Ludwig, Brian H. Walker, and C. S. Holling

Humans are dependent upon natural systems for the necessities of life, such as air and water, as well as for resources that are essential to modern societies (Odum 1993). As humans have imposed greater and greater demands upon natural systems, Arrow et al. (1995) and many others have raised concerns about the sustainability of the resource flows from these systems. The purpose of this exposition is to review some theoretical concepts and present specific examples to illustrate the variety of possible behaviors that natural systems may display under exploitation. The concepts stem from our informal understanding of the ideas of stability, sustainability, and resilience, but clarity requires a more detailed classification of behaviors. The examples that we present do not exhaust the supply of possible behaviors, but each is "generic," in the mathematical sense that small changes in parameter values do not change the qualitative behavior of the system. This implies that the qualitative behavior of each example is typical of a whole class of systems.

Equilibrium

A mechanical system is at equilibrium if the forces acting on it are in balance. For example, when a body floats, the force of gravity is balanced by the buoyant force due to displacement of the liquid. The "balance of nature" (Pimm 1991) is an extension of this idea to the natural world. The concept usually refers to steady flows of energy and materials rather than to systems whose components do not change.

Resilience and Stability

We are interested in characterizing natural systems that are resilient, or, in other words, that tend to maintain their integrity when subject to disturbance (Holling 1973). This is related to the idea of stability. The informal concept of stability refers to the tendency of a system to return to a position of equilibrium when disturbed. If a weight is added suddenly to a raft floating on water, the usual response is for the weighted raft to oscillate, but the oscillations gradually decrease in amplitude as the energy of the oscillations is dissipated in waves and, eventually, in heat. The weighted raft will come to rest in a different position than the unweighted raft, but we think of the new configuration as essentially the same as the old one. The system is stable.

If we gradually increase the weight on the raft, eventually the configuration will change. If the weight is hung below the raft, the raft will sink deeper and deeper into the water as more and more displacement is required to balance the higher gravitational force. Eventually, the buoyant force cannot balance the gravitational force and the whole configuration sinks: the system is no longer stable. On the other hand, if the weight is placed on top of the raft, the raft may flip over suddenly and lose the weight and its other contents long before the point at which the system, as a whole, would sink. This sudden loss of stability may be more dangerous than the gradual sinking, because there may be little warning or opportunity to prepare for it. We may think of the raft system as losing its resilience as more weight is placed on it.

Suppose that we accept the "balance of nature" and the steady flows of resources that it implies. As we demand more and more products of natural systems, and as we load these systems with more and more of our waste products, are we likely to experience a gradual loss of stability or a sudden one? In order to clarify such questions, we must refine our terminology. To decide whether a system is stable or not, we must first specify what we mean by a change in configuration or loss of integrity. If we don't care whether the raft flips over when weighted, then there is no problem of sudden loss of stability for the floating raft. We must also specify the types and quantities of disturbances that may affect the system. Suppose that a fixed weight is placed on top of an occupied raft. If the occupants of the raft move about, the raft may float at a slightly different angle, but if they move too far or all at once, the raft may tip. The range of possible movements of the occupants that do not lead to tipping is called the *domain of stability*, or *domain of attraction*, of the upright state. If the amount of the fixed weight is gradually increased, the balance becomes more precarious and, hence, the domain of attraction will shrink. Eventually, the weight becomes large enough so that there is no domain of attraction at all, and the raft will flip over no matter what its occupants do.

The preceding example makes a distinction between the weight loading the raft and the positions of the occupants. If the amount of the weight changes very slowly or not at all, we may think of the "system" as consisting of the raft and weight. The occupants change position relatively quickly, and these changes may be thought of as disturbances of the system. On the other hand, we may adopt a more comprehensive point of view, seeing the raft, the weight, and the occupants as a single system. If the occupants organize themselves to anticipate and correct for external disturbances, then the system may be able to maintain its integrity long enough for them to achieve their objectives. Another possible response to disturbance might be to restructure the raft itself. If it were constructed of several loosely coupled subunits, then excessive weighting or a strong disturbance might flip one part of the system but leave the rest intact. Such a structure might not require as much vigilance to maintain as the single raft, and it might be able to withstand a greater variety of external disturbances. On the other hand, if the bindings that link the subunits become stiff, then the structure may become brittle and, hence, more prone to failure. This simple example illustrates how the notion of resilience of a system depends upon our objectives, the time scale of interest, the character and magnitude of disturbances, the underlying structure of the system, and the sort of control measures that are feasible.

The section that follows presents the main ideas of stability and resilience for simple, one-dimensional prototype systems. Calculations can be done explicitly for these prototype systems, but their qualitative behavior holds for much more complicated examples. The third section in this chapter illustrates the ideas of bistable equilibria, hard loss of stability, hysteresis, and resilience, with a model for the spruce budworm. There are qualitative similarities between the behavior of the budworm model and a variety of ecological systems, particularly lake ecosystems, the Baltic Sea, and boreal forests, although no attempt is made here to provide a formal model for these systems. Such a model is given in the fourth section, for a competitive grazing system. This system has qualitative behavior analogous to a one-dimensional model, and it also exhibits hysteresis and hard loss of stability. An analogous system that involves fire as a regulating process is presented in the fifth section. The latter system also exhibits regular oscillations. The Appendix presents a detailed account of the relationship between return times for a disturbed system and its resilience. There are two conflicting definitions of resilience, which may cause some confusion. The definition of Pimm (1991) applies only to behavior of a linear system or behavior of a nonlinear system in the immediate vicinity of a stable equilibrium where a linear approximation is valid. For Pimm, loss of resilience is due to slow dynamics near a stable equilibrium. The definition of Holling (1973), which we use in this volume, refers to behavior of a nonlinear system near the boundary of a domain of attraction. Loss of resilience, in our sense, is associated with slow dynamics in a region that separates domains of attraction.

A Simple Prototype for Stability and Resilience

In order to understand complicated systems, it is often convenient to consider a simpler system that exhibits the type of behavior of interest. A full theory of the floating raft would require a combination of the theories of hydrodynamics and of rigid body dynamics, but the essential features can be captured in a one-dimensional model. We are mainly concerned with the notion of stability and the fact that the domain of attraction of a stable equilibrium may depend upon slowly varying parameters. These features are present in a one-dimensional system.

Global Stability

The concept of the balance of nature might be taken to imply that the system will maintain its integrity under any sort of perturbation. Such an assumption may be made (often unconsciously) when we make large modifications to natural systems. Our expectation is that things will proceed more or less as before and that the response of the system will be approximately proportional to the perturbation. Such behavior is shown by the simplest linear models. Some might argue that a principle of parsimony dictates that such models be used in the absence of strong evidence to the contrary. The following linear model illustrates the property of global stability, which implies that the system will always return to a certain equilibrium regardless of how far it is displaced from that equilibrium.

Suppose that the dynamics are given by a relation of the form

$$\frac{dx}{dt} = h(\alpha) - x \tag{2.1}$$

where $h(\alpha)$ is a smoothly varying function of an external variable α and x is the quantity of interest. Then $dx/dt = 0$ if $x = h(\alpha)$; the system has a single equilibrium there. This equilibrium is stable, since $dx/dt > 0$ if $x < h(\alpha)$ and $dx/dt < 0$ if $x > h(\alpha)$. These relations imply that the system approaches the equilibrium, no matter what the starting point.

A system such as equation (2.1) cannot fail us or surprise us. It returns to an equilibrium, no matter how far it is displaced, and the position of the equilibrium changes smoothly with the exogenous variable α. Such a system is not suitable for a discussion of possible collapses of natural systems, since such collapses are excluded by assumptions such as equation (2.1). Mathematical theory provides numerous examples of different behavior, and our goal is to investigate their plausibility. Unfortunately, much theory (including most economic theory) has been based upon assumptions analogous to equation (2.1). In particular, "resilience" has been defined by Pimm (1991) in terms of the system equation (2.1), and this very special assumption may mislead the unwary. There is a mathematical theory that shows that systems such as (2.1) are good approximations to general systems with a stable equilibrium, but that theory implies that the approximation holds only in the immediate vicinity of the equilibrium: the approximation is valid only locally. Details are provided in the appendix.

Bifurcation

In order to explore the differences between local and global stability, we must examine nonlinear models, meaning models in which the state variable appears in functions more complicated than linear ones. The following example has three equilibria instead of a single one. Such a complication requires a cubic or more complicated dependence upon the x variable, for example equation (2.2).

$$\frac{dx}{dt} = f(x) = x(x^2 - \alpha) \tag{2.2}$$

Here, α is a parameter or a slowly varying quantity whose dynamics are not of immediate concern. The equilibria of the system are the states where $f(x) = 0$. These are the states where either of the two factors in equation (2.2) vanishes. Hence, they are points where

$$x = 0 \quad \text{or} \quad x^2 = \alpha \tag{2.3}$$

If $\alpha > 0$, then there are three equilibria (equation [2.4]):

$$x = 0, x = \sqrt{\alpha}, \quad \text{or} \quad x = -\sqrt{\alpha} \tag{2.4}$$

If $\alpha \leq 0$, then there is only the single equilibrium at $x = 0$. Such a change in the configuration or stability of equilibria is called a *bifurcation* (Guckenheimer and Holmes 1983). It implies a change in the qualitative behavior of the system. To explore this feature, we must discuss some additional concepts.

Local Stability and Domain of Attraction

In order to determine the stability of equilibria, it suffices to examine the sign of the velocity of x. For example, if $\alpha < 0$, the second factor in equation (2.2) is always positive and, hence, $dx/dt > 0$ if $x > 0$, and $dx/dt < 0$ if $x < 0$. In this case, the system always moves away from the state where $x = 0$. We conclude that the equilibrium at $x = 0$ is unstable if $\alpha < 0$. On the other hand, if $\alpha > 0$, then dx/dt changes sign at three places:

$$\frac{dx}{dt} > 0 \text{ if } x > \sqrt{\alpha} \qquad (2.5)$$

$$\frac{dx}{dt} < 0 \text{ if } 0 < x < \sqrt{\alpha} \qquad (2.6)$$

$$\frac{dx}{dt} > 0 \text{ if } -\sqrt{\alpha} < x < 0 \qquad (2.7)$$

$$\frac{dx}{dt} < 0 \text{ if } x < -\sqrt{\alpha} \qquad (2.8)$$

The equilibrium where $x = \sqrt{\alpha}$ is unstable, because the system always moves away from that point if nearby (according to equations [2.5] and [2.6]). Similarly, the equilibrium where $x = -\sqrt{\alpha}$ is unstable (according to equations [2.7] and [2.8]). On the other hand, the equilibrium where $x = 0$ is stable (according to equations [2.6] and [2.7]), because the motion from nearby points is toward that point. However, if the system starts outside the interval $(-\sqrt{\alpha} < x < \sqrt{\alpha})$, it moves away from the equilibrium at $x = 0$. Therefore, the equilibrium at $x = 0$ is locally stable, but not globally stable. The system returns to $x = 0$ if small perturbations are made, but larger perturbations take the system into an unstable domain. The interval $(-\sqrt{\alpha} < x < \sqrt{\alpha})$ is called the *domain of attraction* of the point $x = 0$, because trajectories that start within that interval eventually return to $x = 0$, but not those that start outside.

It is clear that the domain of attraction of the stable equilibrium at $x = 0$ shrinks as α decreases toward zero. The three equilibria collapse into one where $\alpha = 0$, and only a single unstable equilibrium remains when $\alpha < 0$. This information is summarized in figure 2.1. The diagram looks like a branch, and, for this reason, it is called a *bifurcation diagram*. The domain of attraction of the point $x = 0$ is contained within the two curved branches, and there is no other domain of attraction.

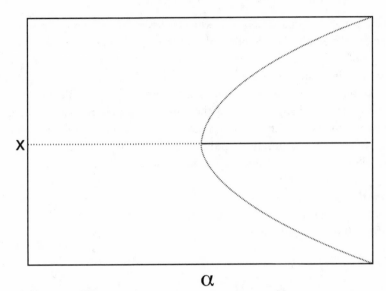

Figure 2.1. The parameter α is plotted on the horizontal axis and the corresponding equilibria in x for equation (2.2) are plotted on the vertical axis.

Disturbances and Slow Parameter Changes

We have seen that if $\alpha > 0$ then this system approaches the stable equilibrium at $x = 0$ if it is started within the domain of attraction. If we envisage disturbances that displace the system a distance x_1 from the stable equilibrium, they will not affect the integrity of the system (its tendency to return to the 0 state) as long as $x_1^2 < \alpha$. Now, if we allow the parameter α to decrease slowly toward x_1^2, the system will take longer and longer to return to the state $x = 0$ when x is displaced to x_1 because motion is very slow near $x = \sqrt{\alpha}$, and a disturbance of magnitude x_1 may take the system into the region of slow dynamics. We may think of the decrease in α as causing a loss of resilience, because the integrity of the system is threatened more and more by disturbances of a given magnitude. A symptom of loss of resilience may be that it takes longer and longer to return to the vicinity of $x = 0$ after disturbance. The connection between return times and resilience is not completely straightforward; we address it in some detail in the appendix.

Two Domains of Attraction

The preceding system is not a believable model for natural systems, because it predicts that the state variable may approach infinity under some circumstances.

A more plausible scenario is one in which the system may change from having a single stable equilibrium to one with two stable equilibria. We next consider a number of such "bistable" systems.

We obtain a simple prototype for such systems by changing the sign of dx/dt in equation (2.2). If the direction of time is reversed, the stable and unstable equilibria are interchanged. If $\alpha < 0$, then the single equilibrium at $x = 0$ is globally stable: the system always returns to that equilibrium no matter where it starts. Instead of two unstable equilibria when $\alpha > 0$ (as in the former case), there will be two stable equilibria. The new system is

$$\frac{dx}{dt} = -f(x) = -x(x^2 - \alpha) \tag{2.9}$$

If $\alpha > 0$ for this system, we have

$$\frac{dx}{dt} < 0 \text{ if } x > \sqrt{\alpha} \tag{2.10}$$

$$\frac{dx}{dt} > 0 \text{ if } 0 < x < \sqrt{\alpha} \tag{2.11}$$

$$\frac{dx}{dt} < 0 \text{ if } -\sqrt{\alpha} < x < 0 \tag{2.12}$$

$$\frac{dx}{dt} > 0 \text{ if } x < -\sqrt{\alpha} \tag{2.13}$$

If $x > 0$ initially, then x heads toward the equilibrium at $x = \sqrt{\alpha}$, but if $x < 0$ initially, then x heads toward the equilibrium at $x = -\sqrt{\alpha}$. Thus, the domain of attraction of the point $x = \sqrt{\alpha}$ is the positive x-axis, and the domain of attraction of $x = -\sqrt{\alpha}$ is the negative x-axis. Each of the stable equilibria is locally stable, but not globally stable. This system can be flipped from one stable state to another by crossing the unstable line where $x = 0$. Because this line separates the two domains of attraction, it is called a *separatrix*. The equilibria in x are plotted in figure 2.2.

Increasing evidence has accumulated for the existence of multistable states in nature: coral reefs (Done 1992; Hughes 1994; McClanahan et al. 1996), African rangelands (Dublin et al. 1990), shallow lakes (Schindler 1990; Carpenter and Leavitt 1991; Scheffer et al. 1993; Carpenter et al. 1999), kelp forests (Estes and Duggins 1995), and grasslands (D'Antonio and Vitousek 1992; Zimov et al. 1995).

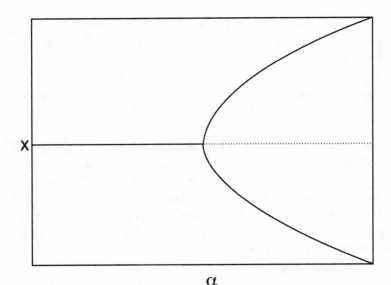

Figure 2.2. Equilibria in *x* for equation (2.9) are plotted against α, as in figure. 2.1.

Disturbances and Slow Parameter Changes

The bifurcation diagram in figure 2.2 implies a great deal about the response of the system to disturbance. If $\alpha < 0$, then the system will return to the stable equilibrium at $x = 0$ no matter how large the disturbance—there is nowhere else for it to go. However, if $\alpha > 0$, and the system starts near the lower branch, it will tend to return there if displaced by a small amount. As α decreases toward zero, the distance between the stable equilibria and the unstable one decreases. Hence, disturbances of a given magnitude take the system closer and closer to the unstable equilibrium. Dynamics are slow near the unstable equilibrium and, hence, the time to return to the vicinity of the lower branch increases sharply for trajectories that approach the unstable equilibrium. This point about return times can be made precise by a calculation analogous to that given in the appendix.

For a higher level of disturbance and $\alpha > 0$, the system may be moved across the separatrix at $x = 0$ and may approach the upper branch of stable equilibria. Under a random pattern of disturbances, we may expect to see the system spend long periods of time in the vicinity of one or the other of the stable equilibria. Every now and then, the random disturbances may combine and send the system to the other stable equilibrium. The *Allee effect* studied by ecologists provides an example of a bistable system. A population may suffer reduced survival or reproductive success at low numbers. For example, schooling fishes tend to suffer low per capita mortality if their numbers are high enough relative to the capacity of their predators. If such fish are reduced in numbers through fishing pressure or environmental degradation, the population may decline and eventually become extinct locally. On the other hand, a large population may sustain itself over long periods. Dynamics of this sort might explain the occasional flips between dominance of sardine and anchovy as revealed by deposits of their scales off the coast of California. A similar pattern appears in such geophysical features as the polarity of the earth's magnetic field, the ocean circulation involving the Gulf Stream, and climate fluctuations, but some of these fluctuations may be too regular to be completely random. For larger values of α one would expect flips from one equilibrium to the other to be extremely rare. Thus, we may associate an increase in α with an increase in resilience.

Hard Loss of Stability and Hysteresis

The two preceding examples illustrate a so-called *soft loss of stability*. As the exogenous variable changes, the location of the stable equilibria changes smoothly. The state variable may move from one domain of attraction to another, but such changes are slow because dynamics are slow near an unstable equilibrium or a separatrix. The possibility of such behavior would not ordinarily be cause for alarm, because slow dynamics may allow for adjustments to new behavior. There are natural systems, such as outbreaking insect populations, that sometimes show more abrupt changes.

The following model was used by Ludwig et al. (1978) to understand the dynamics of the spruce budworm. The quantity B represents budworm density, measured in larvae per acre. This density is assumed to vary in time according to equation (2.14):

$$\frac{dB}{dt} = r_B B (1 - \frac{B}{K_B}) - \beta \frac{B^2}{\alpha^2 + B^2} \qquad (2.14)$$

where r_B is an intrinsic growth rate at low densities, K_B is a carrying capacity for the budworm in the absence of predation, and the second term in equation (2.14) is a predation rate. The predators are assumed to have a Holling type-III functional response, with a maximum predation rate of β and a half-saturation budworm density of α. This functional form implies that predators have their greatest influence upon dynamics at intermediate ranges of budworm densities. At low densities, the predators search for alternate prey because returns from foraging for budworm are relatively low. At high densities, budworms swamp their predators; thus, the predators have a small per capita effect, just as predators have a small per capita effect on large schools of fish. The parameter α is proportional to a measure of foliage density, because the predators search foliage for the budworms and their response is mediated by the number of budworms per unit of foliage. Hence, α is actually a state variable that generally changes on a slower time scale than that of the budworm. For the moment, we regard α as a constant.

Some algebra supplied in Ludwig et al. (1978) shows that there are either two or four equilibria for the budworm, depending upon the sizes of the dimensionless parameters R and Q, given by relations in equation (2.15).

$$R = \frac{r_B \alpha}{\beta}, Q = \frac{K_B}{\alpha} \tag{2.15}$$

These equilibria satisfy equation (2.16),

$$R\left(1 - \frac{b}{Q}\right) - \frac{b^2}{1 + b^2} = 0 \tag{2.16}$$

where $b = B/\alpha$. The equilibrium $b = 0$ is always unstable, because $db/dt > 0$ if b is small and positive. The highest equilibrium is always stable, because $db/dt < 0$ if b is very large and positive. Thus, if there are only two equilibria, budworm density always moves toward the upper equilibrium. When there are four equilibria, they alternate in stability. A typical case is shown in figure 2.3. If R is between R_1 and R_2, then b may approach either the high equilibrium or the low equilibrium, depending upon whether the starting position of b is above or below the unstable equilibrium, which is the separatrix.

Hard Loss of Stability

Imagine that the parameter α begins at a low value and gradually increases as the forest grows. It turns out that Q does not change with forest growth; hence, figure 2.3 applies. Because R is proportional to α, R will increase. At first (when $R < R_1$), budworm numbers will remain low, since the only stable equilibrium is the low one. Even when R increases beyond R_1 the budworm numbers will remain low, because they lie below the unstable equilibrium, which determines the domain of attraction of the low equilibrium. The stability of the low equilibrium becomes precarious as R approaches R_2, because the domain of attraction shrinks. Finally, at $R = R_2$, the lower two equilibria disappear and budworm density jumps to the high value: an outbreak occurs. This abrupt change

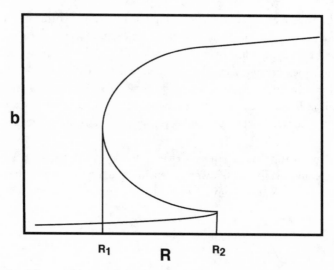

Figure 2.3. The equilibria in *b* for equation (2.16) are plotted against *R* for *Q* = 20.

in the attracting state is called a *hard loss of stability*. It should be contrasted with the soft loss of stability displayed by the system in equation (2.9). In the case of the budworm, once density has reached the high equilibrium there is no easy way to reduce it to the lower equilibrium. If the variable *R* is reduced below R_2, the budworm remains at the high equilibrium. As *R* is further reduced, there is a second hard loss of stability as *R* declines below R_1. In this case, there is a jump down to the low equilibrium, which is not reversed as *R* increases again.

Hysteresis and Cycles

If we now connect the dynamics of the trees and the dynamics of the budworm, a new phenomenon appears. If the system starts with low foliage density and low budworm numbers, the foliage density slowly increases until it surpasses R_2. At this point, an outbreak occurs, as shown previously. High budworm numbers eventually cause death of trees, so *R* begins to decrease when the budworm has an outbreak. Budworm numbers remain high even though *R* declines, because budworm density lies above the separatrix. As *R* continues to decline to R_1, budworm density declines slowly and then jumps to a low value when *R* decreases below R_1. The different paths followed by the total system for increasing versus decreasing *R* constitute the *hysteresis effect*. The combination of budworm and forest dynamics produces stable cycles with long periods. Such stable cycles that are maintained through alternations of rapid transitions and slow changes are called *relaxation oscillations*. They are common in many physical, chemical, and physiological systems (Edelstein-Keshet 1988).

Disturbances and Resilience

If the objective of management is to keep budworm numbers and foliage damage low, the loss of stability as R increases beyond R_2 may be regarded as a loss of resilience. This model suggests that small disturbances near the lower stable equilibrium may exhibit long return times if they approach the unstable branch. Such long return times may be a useful diagnostic indicator. However, because R increases as trees grow, a loss of stability accompanied by a budworm outbreak seems inevitable.

We may adopt a different perspective and regard periodic budworm outbreaks as part of a stable system that renews the forest from time to time. Indeed, systems analogous to the budworm-forest system frequently appear as stable oscillators. The advantage of such oscillators is that they continue to oscillate more or less with the same frequency and amplitude under a wide variety of disturbances. Hence, physiological oscillators are important in maintaining integrity of the organism, which is another kind of resilience. According to this perspective, an attempt to halt the oscillations may lead to a disastrous breakdown in the long term. Will human interventions to increase productivity in natural systems suffer a similar fate?

Lake Dynamics

Carpenter and Cottingham (chapter 3, and 1997) have discussed the applicability of these ideas to lake ecosystems. They characterize lake dynamics as either "normal" or "pathological." Normal lakes have high numbers of game fish, effective grazing upon phytoplankton, and low incidence of algal blooms. The normal system maintains its integrity when subjected to perturbations such as phosphorus pulses, because phosphorus moves rapidly into the higher trophic levels (Carpenter and Kitchell 1993) and humics constrain algal growth (Jones 1992). However, heavy phosphate loading, removal of macrophytes, overfishing, and removal of wetlands and riparian vegetation may lead to the pathological state in which there are few game fish, less grazing, no macrophytes, and extensive and frequent algal blooms (Harper 1992). This may be a rapid transition and it is not easily reversed (National Research Council 1992).

This situation appears to fit the definition of a hard loss of stability, because the change is rapid and large, and is sometimes not reversed even if phosphate loads are decreased (National Research Council 1992). One may say that the normal lake is resilient because it maintains its integrity under perturbation, but resilience is lost as phosphate loading and other stresses are increased. If critical levels of phosphate and other environmental variables could be identified, we might attempt to measure resilience in terms of the difference from the critical levels (Vollenweider 1976). Perhaps the question of whether or not the lake ecosystem fits our definitions may be answered by statistical analysis of long-term data.

The Baltic Sea

Jansson and Jansson (chapter 4) and Jansson and Velner (1995) describe the Baltic system in terms that show many similarities to the lake system. The Baltic Sea is partially enclosed and, consequently, has a residence time of water on the order of twenty years. Algae form the base of a diverse food web, with higher trophic levels occupied by commercially important species such as herring, flounder, pike, and perch. There may be long periods when there is weak vertical mixing of the water column due to lack of inflow from the North Sea. During such periods, oxygen levels at greater depths may be very low and sulphur bacteria may predominate. The latter put large quantities of phosphorus into solution, which then upwell and cause plankton blooms.

In historical times, the Baltic has experienced several extended anoxic periods, but the system had not been permanently altered. Since the industrial revolution, the Baltic has been loaded with increasing amounts of phosphorus, and there are indications of a change of configuration to a detritus-based system. This would imply more turbid water and a fish community consisting mainly of sluggish species such as bream, roach, and ruffe, which are much less valuable than those previously listed. The possibility exists that the Baltic might reach a point at which even reducing phosphate inputs might not return the system to its earlier, more desirable state. Such a turn of events would correspond to a hard loss of stability, analogous to the behavior of the lake ecosystem. Unfortunately, there are no replicates of the Baltic system; hence, we have only analogies to guide action. A purposeful demonstration that the Baltic is actually capable of a sudden change corresponding to a hard loss of stability is unthinkable as an experiment. Nevertheless, it may be occurring as a result of human negligence. The earlier ability of the Baltic to recover from anoxic periods may correspond to resilience, but we cannot be sure whether this resilience is being lost.

The Boreal Forest

Carpenter et al. (1978) characterize the boreal forest as a system with relatively few species and complicated interactions and dynamics. In upland regions, forests dominated by aspen and birch alternate with forests dominated by spruce and fir. Browsing by moose over a period of twenty to forty years can convert an aspen stand into one dominated by conifers. As stands of conifers mature, they become increasingly favorable for reproduction of the spruce budworm. Eventually, outbreaks occur and portions of the system are converted into early successional aspen. The budworm outbreak corresponds to a hard loss of stability, and the combined upland system undergoes stable, long-period oscillations analogous to those described previously.

As the upland regions undergo these oscillations, the valley bottoms alternate between flooded plains and moist meadows. The flooded state is maintained by beavers, which cut aspen bordering streams for food and dam the streams to create ponds. When the supply of aspen is insufficient, the beavers abandon their dams, the dams break, and the ponds are soon replaced by meadows. This relatively rapid change, a consequence of decreasing supply of aspen, may be thought of as a hard loss of stability. The upland and lowland cycles tend to entrain each other because of the interaction between beavers and aspen. Fires also play a role in synchronizing cycles over large spatial areas, because conifers killed by the spruce budworm provide an abundance of fuel.

Although this system undergoes large alterations, sometimes very quickly, it may be thought of as resilient, maintaining its character over many centuries. Conditions at any given site may change abruptly, but the system is usually a mosaic of patches at differing stages of the cycle. When considered as a whole, it maintains considerable diversity.

A Competitive Grazing System

In this section, we describe a natural system that may be bistable. Competition between grasses and woody vegetation in a semi-arid environment is described in chapter 7 and in Walker et al. (1981). Suppose that either the grass or the woody vegetation has an advantage when at high densities relative to the other. In such a case, the system has stable equilibria that correspond to high levels of grass and woody vegetation, respectively. The competition is also influenced by the stocking rate of cattle, which consume grass but not woody vegetation. We shall regard the two plant forms as the dynamic state variables and the stocking rate as a slowly varying parameter.

Imagine starting with high levels of grass and low levels of woody vegetation.

At low levels of stocking, there is only a small difference from the ungrazed system: if the system starts out with grass dominant, grass will continue to dominate. As stocking increases, the competition may favor woody vegetation. Eventually, there may be a collapse of the grass, and woody vegetation will dominate. Thus, the effect of grazing is to move the system from a state in which grass dominates to one in which woody vegetation dominates. Even when grazing pressure is relaxed, there may be little change in composition, because of the advantage enjoyed by woody vegetation over grass when the former is dominant. The effect of grazing is to move the system into the domain of attraction of woody vegetation for the ungrazed system.

If one plots grass density versus the stocking level, the behavior may appear to be inexplicable: the grass level declines as grazing increases but does not return to former levels when grazing returns to its former level. The apparent paradox is resolved if we realize that the density of grass depends not only on the stocking level, but also on competition with woody vegetation. These phenomena may be illustrated by a modification of the Lotka-Volterra competition model.

Let g represent the density of grass, and let w represent the density of woody vegetation. The rate of change of grass density is assumed to be represented in equation (2.17).

$$\frac{dg}{dt} = r_g g(1 - s - c_{gg} g - c_{wg} w) \tag{2.17}$$

In equation (2.17), r_g is a growth rate, and c_{gg} and c_{wg} are competition coefficients. The parameter s is determined by the stocking rate of cattle. The rate of increase of the woody vegetation is assumed to be represented by equation (2.18).

$$\frac{dw}{dt} = r_w[a + w(1 - c_{gw} g - c_{ww} w)] \tag{2.18}$$

In equation (2.18), r_w is a growth rate, c_{gw} and c_{ww} are competition coefficients, and a is a source term. In the illustrations that follow, the parameters were chosen as indicated in equation (2.19).

$$r_w = 1, \ r_g = 1.5, \ c_{gg} = .7, \ c_{wg} = 1, \ c_{gw} = 2, \ \alpha = .03, \ c_{ww} = 1 + a \tag{2.19}$$

The case of light grazing corresponds to $s = 1/10$.

In order to understand the behavior of this system, it is helpful to plot some curves in the g, w plane, as in figure 2.4. In the phase plane, the direction and

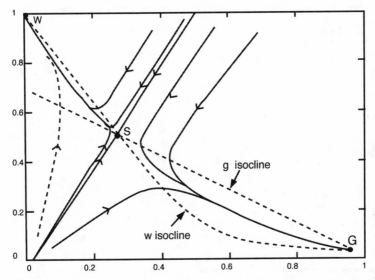

Figure 2.4. The phase plane for grass and trees derived from equations (2.17)–(2.19).

speed of change of the system are given by the vector (dg/dt, dw/dt). This vector is vertical on the curve where $dg/dt = 0$ (the null g isocline), and it is horizontal on the curve where $dw/dt = 0$ (the null w isocline). As can be seen, $dg/dt = 0$, where either $g = 0$, or the relationship holds in equation (2.20):

$$c_{gg}g + c_{wg}w = 1 - s \qquad (2.20)$$

This locus is a straight line, and it shifts to the left as s increases. The null w isocline is a hyperbola according to equation (2.18). One of the asymptotes is the w- axis, and the locus passes through the point $g = 0$, $w = 1$, labeled "W." This locus is independent of the stocking parameter s. Figure 2.4 shows in detail how a system may approach more than one steady state, depending upon the starting conditions. Although this system is much more complicated than equation (2.9), its qualitative behavior is the same if $\alpha > 0$. This illustrates how the very simple one-dimensional models may, nevertheless, be a valuable heuristic guide.

We now turn to the effect of increased stocking. The effect of an increase in s is to shift the null g-isocline down and to the left. Hence, the points S and G will approach each other along the w null isocline. Because the separatrix passes through the point S, the domain of attraction of G must shrink, whereas the domain of attraction of W will expand. Qualitatively, the representation of fig-

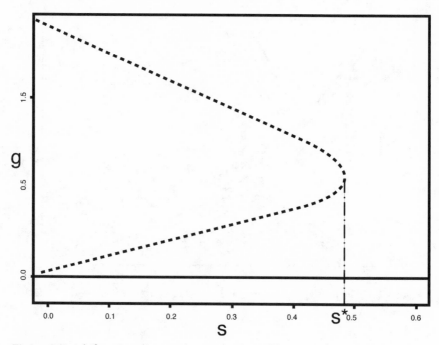

Figure 2.5. A bifurcation diagram showing grass equilibria as a function of the grazing parameter s for the system as described in equations (2.17)–(2.19).

ure 2.4 still holds. For a still-higher value, $s = s^*$ the points S and G coincide. The values of g corresponding to the roots S and G are shown in figure 2.5. A similar diagram could be drawn to show the corresponding values of w. This is a bifurcation diagram analogous to figure 2.3. If the stocking rate changes slowly, we may expect the grass density to be given by the upper curve in figure 2.5. However, if $s > s^*$, the grass density must crash, since there is no stable equilibrium with a nonzero grass density. This is a hard loss of stability, and indeed there is a striking similarity between figures 2.5 and 2.3.

For values of $s > s^*$, the qualitative form of figure 2.6 applies. The only equilibrium is point W, and all trajectories approach W. The domain of attraction of W is the whole quadrant, where $g > 0$ and $w > 0$.

We may imagine the system beginning with the stocking rate $s = 0$. According to figure 2.5, the unstable equilibrium S and the stable equilibrium W are very close together. Hence, the separatrix in a figure analogous to figure 2.4 is in the extreme upper left corner: virtually any initial combination of g and w will lead to a high density of g, given by the upper branch in figure 2.5, with a cor-

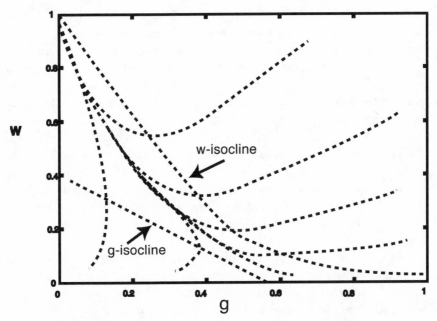

Figure 2.6. The phase plane for grass and trees for the system in equations (2.17)–(2.19), with $S = 0.6$.

respondingly small density of w. Now, if s is increased slowly, the density of g will move downward along the upper branch until the bifurcation point s^* is reached. Beyond that point, the grass density must crash, and woody plants will dominate. Even if grazing pressure is withdrawn ($s = 0$), the grass cannot recover because it will be to the left of the separatrix. This is the hysteresis effect.

Fire in a Savanna System

The preceding model does not describe most savanna systems for two reasons: (1) Generally, neither grass nor woody vegetation can completely exclude the other. Instead, there is a single stable equilibrium for the system (which changes over time, depending upon rainfall, grazing, and fire) where grass and trees coexist. (2) Woody vegetation cannot exclude grass indefinitely: after a long period of low grazing, the grass may return. This is due to a combination of two effects. First, the older woody vegetation may die and leave gaps that may be colonized by grass, and then fire gets into the system. Second, woody vegetation dies back very quickly in dry years but recovers only slowly in wet years—too slowly to use all the water. Grass, on the other hand, can increase ten-fold in a season, quickly enough to fully use all of the available water. Because of this, the combination of wet and dry years keeps woody vegetation at lower levels than the average rain would sustain and permits grass to remain in the system in significant amounts.

When viewed at a long time scale, a brief period of grazing may cause a rapid collapse of the grass followed by a slow recovery. When viewed at a short time scale, there appears to be an equilibrium with high woody vegetation. However, when this system is viewed over a longer time scale, it is apparent that the system merely spends a long time in this state. This sort of qualitative behavior is analogous to "excitable systems." Such systems are best known as models of the nerve impulse according to the theory of Hodgkin and Huxley as modified by Fitzhugh. The system has a single stable equilibrium, but when perturbed in an appropriate direction it may undergo a very large excursion (firing of the neuron), followed by a long recovery (refractory) period. Details are given in Edelstein-Keshet (1988).

To model aging properly is complicated, but for present purposes it suffices to find a simple system that has the required qualitative behavior. We do not contend that the following model is an accurate representation of the true dynamics. We must keep track of surplus grass that may serve as fuel for fires. Hence, we let gross grass production be given by g_p, as in equation (2.21).

$$g_p = r_g g + a_g \tag{2.21}$$

In equation (2.21) a_g is a source term, and r_g is the rate of grass production available for grazing. The grass not consumed by grazing cattle is potential fuel for fires and is denoted by g_f, as defined in equation (2.22)

$$g_f = g_p \exp(-sg) \tag{2.22}$$

The parameter s determines the proportion of grass consumed by cattle. Now equation (2.17) is replaced by equation (2.23).

$$\frac{dg}{dt} = g_f - g_p(c_{gg} g + c_{wg} w) \tag{2.23}$$

The dynamics of w will be influenced by fires, and the age of trees influences their susceptibility to fire. Let a new variable h denote the product of the woody plant density and the average age of the woody plants. A first approximation yields $dh/dt = w$, but that relationship neglects the influence of fire on the average age. The dynamics of w and h are given by equations (2.24) and (2.25), respectively.

$$\frac{dw}{dt} = r_w w (1 - c_{gw} g - c_{ww} w) - fw + a_w \qquad (2.24)$$

$$\frac{dh}{dt} = w - r_f f h \qquad (2.25)$$

The fire risk f is defined as follows: let the fire potential p be proportional to the available fuel as shown in equation (2.26).

$$p = \begin{cases} c_f g_f & \text{if } h > 20 \\ c_f g_f h/20 & \text{otherwise} \end{cases} \qquad (2.26)$$

We assume that the fire risk f is given by equation (2.27).

$$f = p \frac{h^{\alpha}}{(wa_0)^{\alpha} + h^{\alpha}} \qquad (2.27)$$

In equation (2.27), the parameter a_0 is an age at which the fire risk is half of its maximum, and the parameter α determines the sharpness of the increase of fire risk with age. We have used $\alpha = 9$ to give a sharp increase, and $a_0 = 60$. The remaining parameters are given in equation (2.28).

$$r_g = 5, c_{wg} = .8, c_{gg} = 1, a_g = .02, c_f = .4,$$
$$r_w = 3, c_{gw} = .8, c_{ww} = 1, a_w = .02 \qquad (2.28)$$

The stocking rate s may be chosen as a parameter or control variable. In the following, we chose $s = .5$.

Because there are three state variables in this system, one cannot make a meaningful phase plot. However, if the age variable h is fixed, we may gain an impression of the dynamics. Figure 2.7 shows how a low value of h leads the system to an equilibrium with high w. On the other hand, a higher value of h leads to an equilibrium with much lower w as shown in figure 2.8. Now, if h increases with time, the system will first have high w, then lower w as it ages. Fires decrease the average age of trees as well as their density. Hence, the system gets reset to a state analogous to that shown in figure 2.7 after a fire. The cycling of w with time in the full system, with h changing, is shown in figure 2.9.

If one were to observe this system over a time of five to twenty years, it would appear that woody plants would eventually dominate grasses because of the combination of competition and grazing. Over the next thirty years, however, the effect of fire and aging of trees leads to a collapse of the trees, making way for the next cycle. This example illustrates how the time scale over which we observe the system may have a decisive influence upon our classification of its

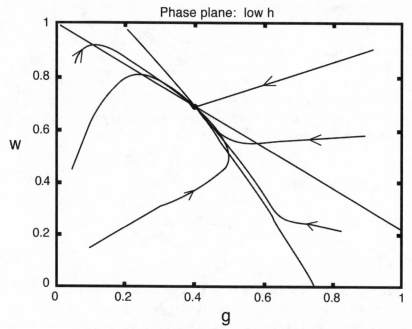

Figure 2.7. The phase plane for grass and trees for the system described in equations (2.21)–(2.28), neglecting the dynamics of the age variable h with a low value of h.

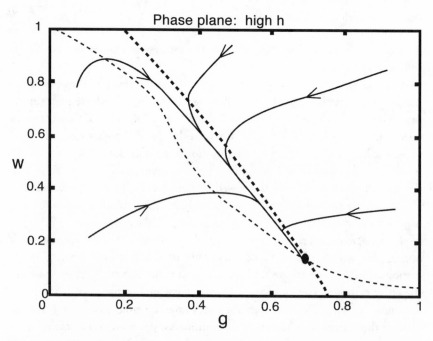

Figure 2.8. The phase plane for grass and trees for the system described in equations (2.21)–(2.28), neglecting the dynamics of the age variable h, with a high value of h.

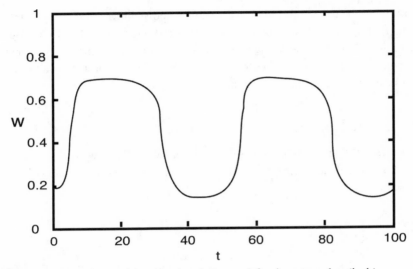

Figure 2.9. Tree density (w) versus time (t, in years) for the system described in equations (2.21)–(2.28), including dynamics for the age variable.

behavior. Unfortunately, the scale over which we are able to observe the system is often much shorter than the scale over which it exhibits its characteristic behavior.

Concluding Remarks

The examples presented here illustrate the complexity of the task facing us as we attempt to clarify the concepts of sustainability and resilience for natural systems. Mathematical theory presents a wealth of possibilities, but the limitations of our understanding and the data available make it difficult to distinguish among them.

Here, we have proceeded from simple conceptual models, such as the raft analogy, to simple but abstract mathematical models, to ecosystem analogues, and finally to a fairly detailed model of a savanna system. In no case can these analogies be considered to be complete, nor is our knowledge of the systems detailed enough to support a full-blown model with statistical justification. Perhaps one might conclude that all this is merely speculation, unworthy of serious attention. On the other hand, prudent decision making requires that we take account of a variety of plausible hypotheses about the responses to our actions. The examples presented here do not encourage complacency about the ability of natural systems to support us and our habits in the lavish fashion we have enjoyed in the past. If we refuse to contemplate possibilities that are only dimly perceived, we may miss the opportunity to learn about the world and adapt our behavior accordingly. If we insist that the simplest and most convenient hypotheses have priority when choosing actions, we run the risk of allowing sloth and pride to bring us to disaster.

Acknowledgments

This research was the product of a Resilience Network Planning Workshop supported by the Beijer Institute, a Division of the Royal Swedish Academy of Sciences.

Appendix
Return Times and Resilience

It is important to distinguish between behavior near a stable equilibrium and behavior near the boundary of a domain of attraction, which is an unstable equilibrium or separatrix. As discussed in the second section of the chapter, the long return times associated with a loss of resilience are caused by slow dynamics near the unstable equilibrium, not by slow dynamics near the stable equilibrium point. Unfortunately, there are two conflicting definitions of resilience and consequent confusion about the connection between resilience and return times.

Pimm (1991 p. 13) defines resilience as "how fast a variable that has been displaced from equilibrium returns to it. Resilience could be estimated by a return time, the amount of time taken for the displacement to decay to some specified fraction of its initial value." Pimm (1991 p. 33) describes return to equilibrium by the equation (A2.1).

$$X_t - X^* = (X_O - X^*)e^{-kt} \tag{A2.1}$$

In equation (A2.1), X_t is the population density at time t, X_o is the initial population density, and X^* is the equilibrium density. The differential equation for X_t that corresponds to this formula is given in equation (A2.2).

$$\frac{dX_t}{dt} = -k(X_t - X^*) \tag{A2.2}$$

A similar model with discrete time could be given instead, but that would not alter the following argument. If we measure displacement from X^* by x, then x satisfies equation (A2.3), which is equivalent to equation (2.1) if $h(\alpha) - x$ is replaced by x.

$$\frac{dx}{dt} = -kx \qquad (A2.3)$$

Strictly speaking, Pimm's definition depends upon this simplicity, because the amount of time required for x to decay to some specified fraction of its initial value is only constant if the model (A2.1) is used. In fact, if the initial displacement is x_0 and the fraction is $p < 1$, then (A2.1) implies the relationships in (A2.4).

$$x_1 = px_0 = x_0 \exp(-kt_r) \qquad (A2.4)$$

From (A2.4), we conclude that the return time t_r is given by equation (A2.5).

$$t_r = \frac{1}{k} \log \frac{1}{p} \qquad (A2.5)$$

The remarkable feature is that the magnitude x_0 does not appear in this formula. This is a feature of this model only, as we shall see below. In more general circumstances, such a result can be expected to hold only in the limit as x_0 approaches 0. Such results are called "local." As pointed in this chapter, a common error is to extrapolate local results to global ones. In the present context, it amounts to replacing a complicated function by a linear approximation. Such approximations are certainly easy to work with, but they may miss essential features of the dynamics. In fact, failure to recognize the distinction between local stability and global stability can lead to unwarranted optimism about the likely consequences of interventions in natural systems. If we think that stability to small perturbations necessarily implies stability to large perturbations, then precautions are never required.

In order to distinguish behavior near the equilibrium at $x = 0$ from behavior near an unstable equilibrium, we must use a model with more parameters than those of (2.2), such as equation (A2.6).

$$\frac{dx}{dt} = f_1(x) = \frac{x(x^2 - \alpha)}{x^2\left[\frac{2}{k_1} - \frac{1}{k}\right] + \frac{\alpha}{k}} \qquad (A2.6)$$

Equation (A2.6) leads to an especially simple equation for the return time: the time to reach a position x_1 starting at x_0 is given by equation (A2.7).

$$t_r = \int_{x_0}^{x_1} dt = \int_{x_0}^{x_1} \frac{dx}{f_1(x)} \tag{A2.7}$$

The form for $f_1(x)$ was chosen so that equation (A2.8) can be verified algebraically.

$$\frac{1}{f_1(x)} = \frac{-1}{kx} + \frac{1}{k_1(x - \sqrt{\alpha})} + \frac{1}{k_1(x + \sqrt{\alpha})} \tag{A2.8}$$

In view of (A2.7) and (A2.8), equation (A2.9) emerges.

$$t_r = \frac{1}{k} \log \frac{x_0}{x_1} + \frac{1}{k_1} \log \frac{x_1 - \sqrt{\alpha}}{x_0 - \sqrt{\alpha}} + \frac{1}{k_1} \log \frac{x_1 + \sqrt{\alpha}}{x_0 + \sqrt{\alpha}} \tag{A2.9}$$

Now, if we replace x_1 by px_0, (A2.9) becomes equation (A2.10).

$$t_r = \frac{1}{k} \log \frac{1}{p} + \frac{1}{k_1} \log \frac{1}{p_1} + \frac{1}{k_1} \log \frac{1}{p_2} \tag{A2.10}$$

Here p_1 and p_2 are given in equations (A2.11) and (A2.12), respectively.

$$p_1 = \frac{x_0 - \sqrt{\alpha}}{px_0 - \sqrt{\alpha}} \tag{A2.11}$$

$$p_2 = \frac{x_0 + \sqrt{\alpha}}{px_0 + \sqrt{\alpha}} \tag{A2.12}$$

If the last two terms in equation (A2.10) are omitted, this result is identical to Pimm's assumption (A2.1). Our more complicated dynamical assumption (A2.6) is the analogue of Pimm's assumption if there are three equilibria. Under what conditions does (A.10) imply large return times? The first term, which corresponds to Pimm's model, implies a long return time if the ratio $p = x_1/x_0$ is small or if k is small. In Pimm's discussion, p is a parameter that describes a probe or observation of the system. Ordinarily, p is fixed, and the return time provides an estimate for k.

The second term in (A2.10) implies a long return time if p_1 is small or k_1 is small. Our previous discussion was concerned with a possibly variable α and disturbances that might take the system near an unstable equilibrium. That corresponds to x_0 near $\sqrt{\alpha}$ or x_0 near $-\sqrt{\alpha}$. In such a case, t_r will be large even if the parameter k is large. That is, return times may be long, even for systems that show very rapid return when close to the stable equilibrium. According to this point of view, long return times may be diagnostic for a small α or for disturbances that are large enough to take the system near an unstable equilibrium.

They may also correspond to weak repulsion from the unstable equilibrium, or, in other words, small k_1. If a disturbance takes the system beyond the unstable equilibrium, there is no return at all.

In summary, according to Pimm (1991) and according to us, long return times may be diagnostic for a loss of resilience, but the meanings of the terms are quite different in the two cases. Pimm is concerned with behavior near a stable equilibrium. In that case, a long return time for a given displacement from the equilibrium indicates a small coefficient k or, equivalently, a small derivative of $\log x$. We are concerned with behavior of a system with two or three equilibria, one of which is stable. Resilience describes the tendency of the system to return to its stable equilibrium. A long return time is due to disturbances that bring the system near an unstable equilibrium, or possibly to a weak repulsion from an unstable equilibrium.

Literature Cited

Arrow, K., B. Bolin, R. Costanza, P. Dasgupta, C. Folke, C. S. Holling, B.-O. Jansson, S. Levin, K.-G. Mäler, C. Perrings, and D. Pimentel. 1995. Economic growth, carrying capacity, and the environment. *Science* 268:520–521.

Carpenter, S. R., and K. Cottingham. 1997. Resilience and the restoration of lakes. *Conservation Ecology* 1:1.

Carpenter, S. R., and J. F. Kitchell, eds.. 1993. *The trophic cascade in lakes*. Cambridge: Cambridge University Press.

Carpenter, S. R., and P. R. Leavitt. 1991. Temporal variation in a paleolimnological record arising from a trophic cascade. *Ecology* 72:277–285.

Carpenter, S. R., D. Ludwig, and W. A. Brock. 1999. Management of eutrophication for lakes subject to potentially irreversible change. *Ecological Applications* 9(3):751–771.

D'Antonio, C. M., and P. M. Vitousek. 1992. Biological invasions by exotic grasses, the grass-fire cycle, and global change. *Annual Review of Ecology and Systematics* 23:63–87.

Done, T. J. 1992. Phase shifts in coral reef communities and their ecological significance. *Hydrobiology* 247:121–132.

Dublin, H. T., A. R. E. Sinclair, and J. McGlade. 1990. Elephants and fire as causes of multiple stable states in the Serengeti-Mara woodlands. *Journal of Animal Ecology* 59:1147–1164.

Edelstein-Keshet, L. 1988. *Mathematical models in biology*. New York: Random House.

Estes, J. A., and D. Duggins. 1995. Sea otters and kelp forests in Alaska: Generality and variation in a community ecological paradigm. *Ecological Monographs* 65:75–100.

Guckenheimer, J., and P. Holmes. 1983. *Nonlinear oscillations, dynamical systems, and bifurcations of vector fields*. New York: Springer-Verlag.

Harper, D. 1992. *Eutrophication of freshwaters*. London: Chapman and Hall.

Holling, C. S. 1973. Resilience and stability of ecological systems. *Annual Review of Ecology and Systematics* 4:1–23.

Hughes, T. P. 1994. Catastrophes, phase shifts, and large-scale degradation of a Caribbean coral reef. *Science* 265:1547–1551.

Jansson, B.-O., and H. Velner. 1995. The Baltic: The sea of surprises. Pp. 292–372 in

Barriers and bridges to the renewal of ecosystems and institutions, edited by L. H. Gunderson, C. S. Holling, and S. S. Light. New York: Columbia University Press.

Jones, R. I. 1992. The influence of humic substances on lacustrine planktonic food chains. *Hydrobiologia* 229:73–91.

Ludwig, D., D. D. Jones, and C. S. Holling. 1978. Qualitative analysis of insect outbreak systems: Spruce budworm and forest. *Journal of Animal Ecology* 47:315–332.

McClanahan, T. R., A. T. Kamukuru, N. A. Muthiga, M. Gilagabher Yebio, and D. Obura. 1996. Effect of sea urchin reductions on algae, coral, and fish populations. *Conservation Biology* 10:136–154.

National Research Council. 1992. *Restoration of aquatic ecosystems: Science, technology, and public policy.* Washington, D.C.: National Academy Press.

Odum, E. P. 1993. *Ecology and our endangered life support systems.* 2nd ed. Sunderland, Mass.: Sinauer.

Pimm, S. L. 1991. *The balance of nature?* Chicago: University of Chicago Press.

Scheffer, M., S. H. Hosper, M.-L. Meijer, B. Moss, and E. Jeppesen. 1993. Alternative equilibria in shallow lakes. *Trends in Ecology and Evolution* 8:275–279.

Schindler, D. W. 1990. Experimental perturbations of whole lakes as tests of hypotheses concerning ecosystem structure and function. *Oikos* 57:25–41.

Vollenweider, R. A. 1976. Advances in defining critical loading levels for phosphorus in lake eutrophication. *Memorie dell'Istituto Italiano di Idrobiologia* 33:53–83.

Walker, B. H., D. Ludwig, C. S. Holling, and R. M. Peterman. 1981. Stability of semiarid savanna grazing systems. *Journal of Ecology* 69:473–498.

Zimov, S. A., V. I. Churorynin, A. P. Oreshko, F. S. Chapin III, J. F. Reynolds, and M. C. Chapin. 1995. Steppe-tundra transition: A herbivore-driven biome shift at the end of the Pleistocene. *American Naturalist* 146:765–794.

Walker, B.H. 1992. Biological diversity and ecological redundancy. *Conservation Biology* 6:18–23.

———. 1995. Conserving biological diversity through ecosystem resilience. *Conservation Biology* 9:747–752.

Walker, B.H., A. Kinzig, and J. Langridge. 1999. Plant attribute diversity, resilience and ecosystem function: The nature and significance of dominant and minor species. *Ecosystems* 2:95–113.

Walters, C. J. 1986. *Adaptive management of renewable resources.* New York: McGraw Hill.

Weaver, J., P. C. Paquet, and L. Ruggiero. 1996. Resilience and conservation of large carnivores in the Rocky Mountains. *Conservation Biology* 10:964–976.

Young, M., and B. J. McCay. 1995. Building equity, stewardship and resilience into market-based property rights systems. In *Property rights and the environment,* edited by S. Hanna and M. Munasinghe. Washington, D.C.: Beijer International Institute and World Bank.

Young, M. D. 1992. *Sustainable investment and resource use.* Paris: Parthenon Publishing Group.

PART II
Resilience in Large-Scale Systems

3
Resilience and the Restoration of Lakes

Stephen R. Carpenter and Kathryn L. Cottingham

Lakes provide humans with services that include water for irrigation, drinking, industry, and dilution of pollutants, hydroelectric power, transportation, recreation, fish, and esthetic enjoyment (Postel and Carpenter 1997). These services are impaired by exploitation of lakes and the lands of their catchments (Hasler 1947; Edmondson 1969; Harper 1992). Because human effects on lakes are growing, concern increases that lake ecosystem services are in jeopardy (Naiman et al. 1995). These concerns parallel those for the sustainability of services from many ecosystems and the biosphere itself (Arrow et al. 1995; Levin 1996).

Scientific studies of lake ecosystem processes have increased in spatial and disciplinary scope during the past century. Although many insights have been derived from the view of lakes as bounded systems defined by the land-water interface, limnologists recognize that lakes must be understood in the landscape context of their catchments (Likens 1984; Wetzel 1990). Airborne pollutants and stratospheric ozone depletion connect lakes to perturbations of the global environment (Schindler et al. 1996b). Changes in agriculture, riparian land use, forestry, fossil fuel consumption, and demand for ecosystem services link lakes to much larger social and economic systems (National Research Council 1993; Postel and Carpenter 1997).

Conceptual development has not kept pace with changes in the processes that alter and control lakes. Ecosystem ecology has a rich understanding of the physical, chemical, and biotic processes of lakes and their watersheds. There is a growing appreciation for the processes that explain heterogeneity among lakes in regional landscapes, regional effects of atmospheric deposition of contaminants in lakes, and the potential effects of global climate change on lakes (McKnight et al. 1996). However, we lack a conceptual framework for understanding the interactions of people and lakes. How do social and economic activities affect lakes? How do properties of lakes affect people's behavior toward lakes? What policies or institutions might sustain lakes and the services they provide? Such questions have barely been asked, let alone answered. Yet, sustainable restoration of lakes must address the social and economic, as well as the biotic and chemical, causes of lake degradation.

Concepts that integrate people and lakes must consider the processes that control normal and degraded lakes. Some relevant processes include climate, regional economic and social activity, watershed vegetation and land use, and internal components ranging from fish and macrophytes to phosphorus and humic staining. This chapter describes the processes that control resilience of lakes, the pathology of degraded lakes, and lake restoration. Our goal is to synthesize published information and suggest some patterns and connections that contribute to an understanding of the interactions of lakes and people. The literature we review is drawn mostly from lake districts of North America and western Europe, where the most conspicuous human effects are associated with agriculture, urban development, industry, and recreation. However, our hypotheses about differences in control processes for normal and degraded lakes may apply to many of Earth's lake districts.

Resilience Mechanisms

In this volume, a resilient system is one that tends to maintain a given state when subject to disturbance (Holling 1973; Ludwig et al. 1997). Lakes are routinely disturbed by many kinds of events. Inputs of solar radiation fluctuate from second to second. Pulses of chemical inputs occur with storms (at intervals of days to weeks) and annual snowmelt. Fluctuations in climate affect seasonal phenologies and budgets of heat, water, and nutrients from year to year (McKnight et al. 1996). Exceptionally large cohorts of fishes recruit every few years and restructure the food web for their lifetimes. Fires or other disturbances alter watershed vegetation at intervals of decades to centuries thereby causing a pulse of nutrients to the lake and changing the water budget (through changes in evapotranspiration) as the vegetation regrows. Extreme changes in climate (with cycles of centuries) change hydrologic connections, water level, the shape of the lake, or even cause the lake to disappear for a period of time.

In the normal dynamics of lakes, ecosystem processes are maintained despite moderate and continuous disturbances originating in the lake, its watershed, and its airshed (figure 3.1). This resilience involves several mechanisms, which

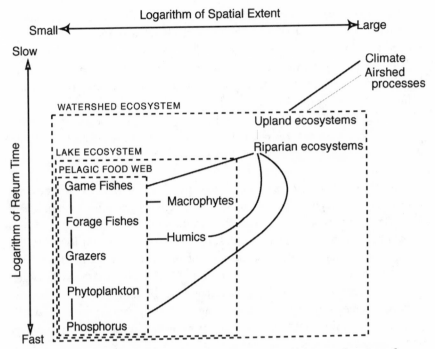

Figure. 3.1. Major interactions in the normal dynamics of lakes as a function of spatial extent (*x*-axis) and return time (*y*-axis).

have different ecosystem components and distinctive spatial locations, spatial extents, and return times.

Riparian forests and grasslands delay or prevent nutrient transport from uplands to streams and lakes (Osborne and Kovacic 1993). Riparian forests are a source of fallen trees that can provide important fish habitat for decades (Maser and Sedell 1994; Christensen et al. 1996).

Wetlands function as vast sponges that delay the transport of water to downstream ecosystems and, thereby, reduce the risk of flooding (National Research Council 1992). Wetlands also modulate nutrient transport from uplands to streams and lakes (Johnston 1991). Wetlands are a major source of humic substances for lakes (Hemond 1990; Wetzel 1992). This complex of organic compounds stains lake water and affects ecosystem metabolism through several mechanisms (Jones 1992; Wetzel 1992). Of particular importance, humic staining suppresses the response of phytoplankton to pulses of nutrient input (Vollenweider 1976). This resilience mechanism involves shading, effects of humics on thermal structure of lakes, and changes in lake metabolism (Carpenter and Pace 1997).

Although phosphorus inputs and recycling establish the potential productivity of lakes (Schindler 1977), predation controls the allocation of phosphorus for production of fish, algal blooms, or other components of the pelagic food web (Carpenter and Kitchell 1993). In the normal dynamics of many lakes, large piscivorous game fishes are keystone predators that structure the food web below them (Kitchell and Carpenter 1993). Such lakes have large-bodied zooplankton grazers that effectively control phytoplankton (Carpenter et al. 1991). When pulses of phosphorus enter these lakes, the nutrient is transferred effectively to higher trophic levels and does not accumulate as algal biomass (Carpenter et al. 1996; Schindler et al. 1996a).

Low or moderate rates of phosphorus input promote low rates of phosphorus recycling through effects on the oxygen content of the water. Conditions of low-to-moderate productivity constrain respiration by bacteria, so that oxygen is not depleted from deeper waters during summer (Cornett and Rigler 1979). Oxygenated conditions decrease the rate of phosphorus recycling from sediments in many lakes (Caraco 1993). If production of the overlying water increases, deep waters can be deoxygenated and phosphorus recycling can increase, thereby further increasing production. Oxygenation of bottom waters prevents this positive feedback and confers resilience in moderately productive and unproductive lakes.

Submersed macrophytes of the littoral zone provide crucial habitat for attached algae, invertebrates, and fishes (Heck and Crowder 1991; Moss 1995). They also modify inputs to lakes from riparian or upstream ecosystems, store substantial amounts of nutrients, and are a source of dissolved organic compounds (Wetzel 1992). Oxygen production by macrophytes and attached algae can decrease the rate of phosphorus release from sediments, and high denitrification rates in littoral vegetation can decrease nitrogen availability (Wetzel 1992).

Collectively, these resilience mechanisms, operating at diverse scales, buffer lake ecosystems against fluctuating inputs. They maintain water quality, fish productivity, and the reliability of other ecosystem services provided to humans.

Measurements and scientific analyses of perturbations or resilience are always tied to particular scales of space and time (O'Neill et al. 1986). In the normal dynamics of lakes, perturbations are relatively brief in duration but may be extensive in space. Examples are chemical or hydrologic fluctuations driven by weather, routine fluctuations of interacting populations, or fires that sweep through the watershed vegetation. Resilience mechanisms that tend to restore the normal dynamics involve longer or larger scales. Examples are food web dynamics that absorb nutrient pulses, wetlands that retain nutrients and release humic substances, or secondary succession of upland forests that stabilizes soils and retains nutrients. These resilience mechanisms can be destroyed by more extreme perturbations. Destruction of the normal resilience mechanisms is accompanied by the rise of new resilience mechanisms and qualitative changes in the ecosystem.

Pathological Dynamics

In normal lake dynamics, the dominant controlling processes are located in the watershed and the lake itself. In the pathological dynamics of lakes, the control shifts to processes with larger spatial extent (figure 3.2). Our use of "pathological" follows that of Leopold (1935): "Regarding society and land collectively as an organism, that organism has suddenly developed pathological symptoms, i.e., self-accelerating rather than self-compensating departures from normal functioning." Although the analogy of ecosystems to organisms now seems dated, many aspects of Leopold's essay seem current. Leopold recognized that both normal and degraded states of ecosystems could be self-sustaining. His concept of recuperative capacity ("capacity, when disturbed, to establish new and stable equilibria between soil, plants, and animals") is similar to the modern concept of resilience. Leopold also recognized that people and nature must be viewed as an integrated system to understand the self-sustaining properties of normal and degraded ecosystems.

Degradation of lakes is a syndrome that involves breakdown of several resilience mechanisms and formation of several new ones (figure 3.2). The most common causes of lake degradation are pollutants from a variety of sources, especially agriculture (National Research Council 1992, 1993). The changes in land use that create and sustain pollutant flows are rooted in economic, demographic, and social changes that link lakes to large-scale human systems.

Eutrophication is probably the best understood type of lake degradation (Hasler 1947; Edmondson 1969; Harper 1992). Agriculture and urban development increase phosphorus transport to lakes. Losses of riparian vegetation and wetlands increase phosphorus flows. Humic inputs decline and humic constraints on phytoplankton become less effective. At the same time, game fish abundance is often reduced by overfishing, so planktivorous forage fishes and bottom-feeding fishes become more abundant. The large zooplanktonic grazers are reduced, and incoming phosphorus accumulates in phytoplankton biomass, especially in blue-green algae. Macrophyte beds decline because of reductions in water clarity and disturbance by bottom-feeding fishes. Loss of crucial habitat (macrophytes, wetlands, fallen trees) leads to further breakdown of the food web. The result is a lake with few piscivorous game fishes, abundant plankton- and bottom-feeding fishes, few large, herbivorous zooplankton, few macro-

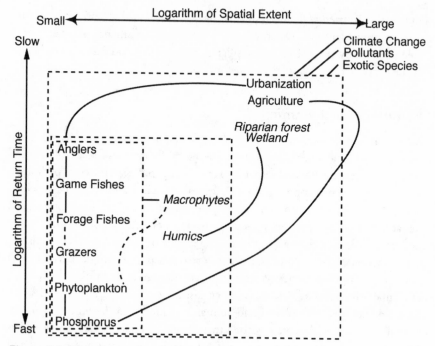

Figure 3.2. Major interactions in the pathological dynamics of lakes as a function of spatial extent (*x*-axis) and return time (*y*-axis).

phytes, dense algal blooms, and risks of anoxia and algal toxins. Toxic pollutants, species invasions, and species extirpations may interact with effects of phosphorus enrichment, habitat loss, and overfishing to exacerbate the degradation of lakes.

The dominant forces driving this degradation are regional to global in extent. Phosphorus pollution from manure and eroding cropland is a principal cause of lake degradation in the United States (National Research Council 1992). Farms import phosphorus, in the form of animal feeds to grow livestock, and as fertilizer to grow crops (National Research Council 1993). Some of the excess phosphorus drains into lakes to cause eutrophication (Daniel et al. 1994). Farmers' decisions about livestock densities and fertilizer applications are linked to regional or global markets for meat, dairy products, and grain.

Urban development, especially in wetland or riparian areas, is another principal cause of lake degradation (National Research Council 1992). Urban development is linked directly to human demography and regional economics. Likens (1992) identifies an important feedback in recreational lake districts called "leapfrog degradation." Riparian homeowners create lawns, remove shoreline vegetation, and harvest fish. Phosphorus inputs from fertilizers and septic systems promote algal growth, and losses of woody habitat and overfishing deplete fish stocks. The lake degrades, becoming less attractive for recreation. Consequently, development shifts to less-disturbed lakes and the cycle is repeated.

Airborne pollutants and exotic species have become global controls of lake dynamics. Mercury, for example, is a widespread toxic pollutant derived from fossil fuels and other sources (Porcella et al. 1995). Under certain conditions, mercury is methylated and bioaccumulates through food chains (Driscoll et al. 1994). Organochlorine compounds are persistent toxic pollutants that affect many lakes. Although the more pernicious organochlorine compounds are no longer manufactured in the United States, significant amounts of some compounds still cycle within lakes and enter lakes from the atmosphere (Rasmussen et al. 1990; Stow et al. 1995). Acid precipitation has several negative impacts on lakes (Schindler et al. 1991), including a surprising interaction with stratospheric ozone depletion (Schindler et al. 1996b). As lakes become more acid, humic concentrations decline and the water becomes more transparent to ultraviolet radiation, exacerbating the impacts on aquatic organisms of increased UV intensities caused by declining ozone concentrations in the upper atmosphere.

Lakes, like islands, can be changed dramatically by species invasions. One spectacular example is the Laurentian Great Lakes, where invading sea lamprey (and overfishing) extirpated piscivorous lake trout thereby allowing populations of another exotic species, the planktivorous alewife, to explode (Christie 1974). These exotics probably entered the Great Lakes through shipping canals. Fish stocking, ballast water, recreational boat traffic, and the aquarium trade have also transported exotics to new lakes. Some, but by no means all, invading species have powerful impacts on lakes, but our abilities to predict the effects of an invader in advance of the invasion are limited (Lodge 1993). Like the chemical changes in lakes, these biotic changes are driven by regional and global alterations of human activity and enterprise.

The pathological dynamics of lakes are also resilient. Ironically, this resilience can counter lake restoration efforts. In numerous cases, regulation of phosphorus in sewage effluents has been offset by increases in agricultural run-off or recycling from sediments (National Research Council 1992). Contaminant concentrations in fishes are stabilized by atmospheric inputs, recycling from sediments, and efficient trophic transfer (Stow et al. 1995). Restoration of piscivorous fishes has been countered by increases in angling (Kitchell 1992). To succeed, lake restoration programs must overcome the resilience of degraded lakes.

Alternative States

Alternative stable states (chapter 2 and Ludwig et al. 1997) may be evident at several levels of watershed-lake systems. Watershed geochemistry, vegetation, and land use strongly affect lakes, and agricultural and urban development of watersheds usually shifts lakes toward eutrophic states. Changes in watershed land use and fisheries management can cause abrupt and massive shifts in limnological conditions.

Many existent feedbacks act to stabilize lake trophic states of oligotrophy (low productivity, low nutrients, and clear water), dystrophy (low productivity and humic-stained water), and eutrophy (Wetzel 1983). These classically recognized states of lakes appear to be consistent with the concept of alternative stable states (Carpenter and Pace 1997). The interaction of humic inputs, phosphorus inputs, and planktonic metabolism can create alternative states of eutrophy and dystrophy under certain conditions. Rapid transitions between states can be caused by perturbations of chemical inputs or food web structure (Carpenter and Pace 1997).

Gradual changes in the planktivory rate can cause abrupt shifts in plankton community structure, grazing, algal biomass, and water clarity. Scheffer (1991) described this process using a model with two alternative states, turbid and clear. Dynamics of a cisco-zooplankton-phytoplankton food web in Lake Mendota, Wisconsin, exhibit abrupt transitions that closely resemble alternative states (Rudstam et al. 1993). The model of Carpenter (1988) also exhibits turbid and clear states, but the dynamics follow an oscillation analogous to the spruce budworm system of Ludwig et al. (1997). In Carpenter's model, the oscillations are generated by a trophic cascade coupled to year classes of an age-structured piscivore population with density-dependent recruitment. Post et al. (1997) describe a case study of the cascading effects of largemouth bass year classes that corroborates the predictions of the model. The model also predicts that long-term fluctuations of phytoplankton should contain a cyclic component, with periods roughly equal to the mean generation time of the dominant pelagic predator in the lake. Carpenter and Leavitt (1991) demonstrated such periodicities in a paleolimnological record.

The dynamics of shallow lakes provide the best limnological example of alternative states (Scheffer et al. 1993). Shallow eutrophic lakes exist in two states: turbid and dominated by phytoplankton, or clear and dominated by macrophytes (usually rooted aquatic plants, but sometimes attached macroalgae). The turbid state involves dense phytoplankton growth driven by nutrient recycling from sediments. Shading by phytoplankton blocks growth of attached plants. The plant-dominated state involves dense growths of attached plants that stabilize sediments (thereby slowing nutrient recycling) and that shelter phytoplankton grazers. The switch to the plant-dominated state can be triggered by a trophic cascade: piscivore stocking and/or planktivore removal to increase grazing and reduce phytoplankton (Jeppesen et al. 1990; van Donk and Gulati 1995; Moss et al. 1996). The switch to the turbid state can be triggered by water-level fluctuations (Blindow et al. 1993) or grazing of macrophytes by fish or birds (van Donk and Gulati 1995). Some lakes have switched states several times, with intervals of years to decades passing between transition events (Blindow et al. 1993).

Resilience and Return Time

Twenty-five years ago, resilience (Holling 1973) and return time (May 1973) were distinct concepts in ecology. Resilience is the capacity of a system to maintain certain structures and functions despite disturbance, and return time is the period required for a system to return to a particular configuration following disturbance. The return time to a stable equilibrium can be used to measure resilience of linear systems, or of nonlinear systems if the perturbations are small (Pimm 1984; DeAngelis 1992; Ludwig et al. 1997). In these situations, changes in return time can be used to measure changes in resilience.

In nonlinear ecosystems subject to large disturbances, however, losses of resilience are often associated with slow dynamics near an unstable point (Ludwig et al. 1997). In these situations, return times near stable points may be irrelevant. To address this issue, Holling (1973) and Ludwig et al. (1997) outline a broad view of resilience that depends on management objectives and options, the time scale of interest, and the type and magnitude of perturbations.

Although return rate near a stable point may miss important features of a system's resilience, empirical estimates of return rate may be useful for comparing responses of different systems to a given perturbation. Such measurements will usually be scale dependent, and there may be more than one possible explanation for an observed difference in return rate.

Return rate is related to nutrient turnover rate in diverse models of lake ecosystems (DeAngelis et al. 1989; Carpenter et al. 1992; DeAngelis 1992; Cottingham and Carpenter 1994). Phosphorus turnover rate is straightforward to measure and provides a means of testing hypotheses about lake ecosystem return rate (Cottingham and Carpenter 1994). These studies were conducted on lakes, not watersheds. Applications of this idea at watershed or regional scales have not yet been explored.

Analyses of experimental lakes have tested hypotheses about resilience and return rate. In lakes, shorter food chains have faster return rates (Carpenter et al. 1992), confirming a conjecture of Pimm and Lawton (1977). These fast return rates are due to rapid phosphorus dynamics of the smaller-bodied fishes that dominate the shorter food chains of lakes (Carpenter et al. 1992; Cottingham and Carpenter 1994). However, the shorter food chains are less capable of assimilating phosphorus pulses than are longer food chains (Carpenter et al. 1992; Schindler et al. 1996b). In the short food chains, phosphorus pulses facilitate growth of grazing-resistant phytoplankton, and phosphorus accumulates as algae. Longer food chains contain larger-bodied grazers that consume a wider range of phytoplankton species and thereby transfer phosphorus to consumers more effectively than the smaller-bodied grazers of shorter food chains (Carpenter and Kitchell 1993). Phosphorus pulses are absorbed by consumers or defecated to sediments in the longer food chains.

Recently, Ives (1995) proposed a method for estimating stochastic return time from noisy time series observations of ecosystems. He suggests that the ratio of variability in population growth rates to variability in population densities (estimated using regression models of a particular form) measures return rate for stochastic systems in a way analogous to return rate for deterministic systems. Ives' insight creates the opportunity to measure and compare stochastic return times for a wide variety of ecosystems, using actual time series data.

Ives' methods also suggest that longer food webs, dominated by large, piscivorous game fishes, are more resilient to phosphorus inputs. Carpenter et al. (1996) measured grazer biomass and chlorophyll for four years in lakes with contrasting food webs that were enriched with phosphorus. Piscivore-dominated lakes contained largemouth bass, few planktivorous fishes, and large-bodied grazers, whereas planktivore-dominated lakes contained no piscivores and abundant planktivorous fishes and small-bodied grazers. The stochastic return rate was calculated for the total grazer biomass and the phytoplankton biomass. Ives' (1995) index of stochastic return rate was 0.89 in an unenriched piscivore-dominated lake, 0.51 in an enriched piscivore-dominated lake, and 0.10 in an enriched planktivore-dominated lake. These calculations show that phosphorus enrichment reduces stochastic return rate, and reduces it far more in the planktivore-dominated lake than in the piscivore-dominated lake.

In summary, resilience to phosphorus inputs in pelagic food webs depends on control of the phosphorus cycle by fishes. In piscivore-dominated lakes, the phosphorus return rate of the entire food web is slow because of the relatively slow phosphorus turnover rates of the fishes, which are a large phosphorus pool. However, by sharply reducing planktivory, the piscivores allow the development of a planktonic subsystem that efficiently converts phosphorus inputs to zooplankton biomass or sediment. Thus, phosphorus inputs do not accumulate as phytoplankton. If the piscivores are removed, planktivorous fishes dominate the food web and the phosphorus return rate of the entire food web increases due to faster phosphorus turnover by the smaller-bodied planktonic fishes. However, the efficient planktonic subsystem collapses and is replaced by a subsystem in which phosphorus inputs accumulate as phytoplankton and symptoms of eutrophication are exacerbated.

Assessment of Resilience

Earth's lake districts contain on the order of 100 million lakes that are greater than 1 hectare in area, and about 1 million lakes greater than 1 square kilometer in area (Wetzel 1990). Lakes are often used as sentinel ecosystems because they collect and integrate regional signals and preserve long-term information in their sediments. However, monitoring lakes at regional to continental scales is a significant challenge. Assessment of many lakes at the landscape scale, over sustained periods of time, requires relatively inexpensive and rapid assays. What indicators are appropriate for large-scale, long-term studies of the regional resilience of lake districts? Resilience derives from partially redundant control processes that act at different scales to mitigate effects of perturbations. In normal lakes, key components are riparian vegetation, wetlands, game fish, and macrophytes. When these are intact, watershed-lake systems can withstand shocks at several scales, such as droughts, floods, forest fires, and recruitment variations. When these control mechanisms are broken down, control shifts to regional economic factors related to farm phosphorus budgets, development, and fishing. Perturbations translate into algal blooms and other symptoms of persistent eutrophication.

Our analysis suggests several potential indicators of a lake's capacity to maintain normal dynamics (figure 3.1). Livestock density in the watershed is a correlate of phosphorus imports (National Research Council 1993). Wetland area per unit lake area is an index of the landscape's capacity to hold water and export humic substances (Wetzel 1990). The proportion of the riparian zone occupied by forest and grassland indicates the potential attenuation of nutrient inputs (Osborne and Kovacic 1993). Lake color relates to humic content (Jones 1992). Slow-to-moderate piscivore growth rates are associated with strong piscivore control of planktivores (Kitchell et al. 1994). Grazer body size correlates with the capacity to suppress algal growth (Carpenter et al. 1991). Partial pressure of carbon dioxide in surface waters may be a sensitive indicator of ecosystem metabolism (Cole et al. 1994). Hypolimnetic oxygen depletion is a symptom of eutrophication and a driver of phosphorus recycling from sediments (Cornett and Rigler 1979; Caraco 1993). All of these indicators can be determined fairly inexpensively from agricultural data and land use records, remote sensing, and sampling to collect surface water, an oxygen profile, a zooplankton haul, fish size distribution, and fish scales for age determination. Calibration of these indicators using extant long-term records and broad regional surveys is a research priority.

Significance of Biodiversity for Resilience

The importance of biodiversity stems from the imbrication of control processes that maintain ecosystem resilience. Each control process may involve many species. The relative importance of a species for a given control process may vary from lake to lake or from time to time. Certain keystone species (Power et al. 1996) and "ecological engineers" (Jones and Lawton 1995) strongly influence ecosystem processes and resilience. Other species depend on the resilience mechanisms for their continued existence but do not have obvious effects on resilience (Walker 1992). Thus, the association of biodiversity with resilience appears complex and variable.

Lake ecosystems can be configured in only a limited number of ways. Certain control processes are repeated in lake after lake, time after time. These regularities make it unnecessary to study each lake as if it were a new, unique system. Instead, we can look for general structures and processes that explain broad patterns.

Conservation of ecosystem structures is marvelously illustrated by Thingvallavatn, a deep, isolated, young lake carved by a glacier from a volcanic rift valley in Iceland (Campbell 1996). The food web of Thingvallavatn is dominated by arctic char. Over the past few thousand years, the chars have diversified from a common ancestor into four varieties: a planktivore, a small benthivore, a large benthivore, and a piscivore. The varieties differ substantially in maximum adult size, habitat use, mouth morphology, and diet. Thus, the ecosystem structure of Thingvallavatn closely resembles that of other lakes with more diverse fish assemblages. The resource polymorphisms of Thingvallavatn's arctic chars suggest that there are only certain limited ways in which lake ecosystems can be structured. This conservation of ecosystem structure resembles the regular clusters of adult body sizes described for terrestrial animal communities (Holling 1992; Holling et al. 1996).

Yet, how resilient is Thingvallavatn? A disturbance that affected arctic char could eliminate several ecosystem processes. Where one taxon controls an ecosystem process, species change and ecosystem change go hand in hand. Acidification of lakes, for example, eliminates a key group of bacteria, thereby blocking the cycle of nitrogen, one of the most important nutrients (Rudd et al. 1988). Would resilience be greater if several species were capable of performing each process? Experimental studies of more diverse lakes suggest that the answer is yes.

The role of species diversity in lake resilience is illustrated by experiments in which lakes were manipulated to various levels of toxic chemical stress or nutrient input. At low levels of toxic stress, changes in species composition are substantial, but changes in ecosystem process rates are negligible (Schindler 1990; Howarth 1991; Frost et al. 1995). Structural change at the species level stabilizes ecosystem process rates, an example of resilience called *functional compensation* by Frost et al. (1995). At more extreme levels of toxic chemical stress, functional compensation is not possible because too many species have been lost; consequently, ecosystem process rates change (Schindler 1990; Frost et al. 1995). Nutrient enrichment, in contrast, simultaneously changes both species composition and ecosystem process rates (Cottingham 1996). In this case, functional compensation allows ecosystem process rates to rapidly track shifts in availability of the limiting nutrient. Phytoplankton species turnover rates are high at all enrichment levels. The availability of a large species pool may facilitate lake response to enrichment in the same way that high diversity facilitates grassland recovery from drought (Tilman 1996). Studies of both toxic stress and nutrient enrichment show that we cannot predict which species will account for functional compensation or other responses to manipulation (Frost et al. 1995; Cottingham 1996). Ecosystem response to a given perturbation depends on only a fraction of the species pool, but the critical species are situation specific and can rarely be anticipated.

In summary, biodiversity confers resilience through compensatory shifts among species capable of performing key control processes. However, species number is not necessarily a good predictor of resilience. The link between biodiversity and resilience depends on the dominant ecosystem control processes, the complement of species capable of contributing to each process, and the susceptibility of these species to a particular ecosystem stress.

Policies for Restoration

Breakdown of lake-watershed systems has been caused by policies that accelerate agricultural phosphorus flows, draining and development of wetlands, removal of riparian vegetation, overfishing, and spread of exotic species. The transformed, degraded state is also resilient. This resilience must be overcome to restore watersheds and lakes. But, according to the U.S. National Research Council (1992), "many so-called lake restoration projects really are only mitigation and management efforts to rid a lake, by whatever means, of some nuisance."

Reactive management, which responds to the symptom of the moment but does not address systemic causes or long-term solutions, characterizes many water quality management programs (Soltero et al. 1992). Reactive management systems are subject to periodic upheavals precipitated by crises that expose the inadequacies of current policy (Gunderson et al. 1995). In the United States, three cycles of water-quality policy can be recognized (Carpenter et al. 1998). Public health concerns that arose in the course of federal water projects led to institutions for water treatment during the 1920s and 1930s. By about 1950, it was clear that lake eutrophication was widespread and unchecked. Management attention focused on point-source phosphorus controls, such as sewage treatment plants. As reductions of point-source phosphorus inputs became more common, it was evident by the 1970s that run-off from farms, construction sites, and urban areas was responsible for widespread degradation of lakes. We are still seeking institutional mechanisms capable of controlling run-off and restoring lakes degraded by pollution from diverse, disaggregated sources.

Sustainable restoration is not reactive; its goal "is to emulate a natural, self-regulating system that is integrated ecologically with the landscape in which it occurs" (National Research Council 1992). In other words, restoration requires shifting resilience mechanisms from those that maintain degraded systems (figure 3.2) to those that maintain more valuable systems (figure 3.1). Resilient restorations will tend to be self-sustaining. For lakes, resilient restorations require restoration of riparian, wetland, and macrophyte vegetation; reduction of phosphorus imports to farms; and reduced harvests of game fish. Ultimately, these changes are linked to social and economic processes at regional to continental scales.

The value of ecosystem services provided by lakes is substantial. Postel and Carpenter (1997) estimate that the total global value of freshwater ecosystem services is several trillion U.S. dollars per year. In the United States, the additional economic benefit of increasing lake water quality to meet acceptable standards for boating, fishing, and swimming is estimated at 31 to 55 billion U.S. dollars per year (Adler and Landman 1993; figures are adjusted to 1995 U.S. dollars). This calculation is an underestimate because it does not include benefits such as flood control, pollution dilution, reduced costs of purifying drinking water, and increased utility of cleaner water for irrigation and industry (Adler and Landman 1993). Degraded lakes represent the loss of substantial economic benefits.

In view of the economic benefits of clean lakes, it seems surprising that lakes continue to degrade and that restoration programs are few. The ecological causes of the problem are understood, and many useful technologies for lake restoration exist (National Research Council 1992). Why is restoration so difficult?

The fundamental problem of lake restoration is an economic mismatch: those who cause the problem do not benefit sufficiently from the remediation. On the other hand, the beneficiaries of lake restoration are not those who caused the degradation. The economic benefits of clean lakes need to be channeled in ways that create incentives for conservation of phosphorus on farms, restoration and maintenance of wetlands and riparian vegetation, and conservation of macrophytes and game fish. Thus far, the United States has not devised social and institutional mechanisms that achieve this fundamental goal.

Scientists can help in understanding this mismatch by shifting their analyses to an appropriately large reference frame. Aquatic scientists now study environments that are fundamentally different from those of the past. Although the basic biological and physical-chemical principles of limnology are unchanged, the dominant controls of lake ecosystems have shifted to much larger spatial scales. Ancestral local controls of lakes are much weakened and subordinated to new controls that involve regional trends in angling and fishing technology, economic and demographic forces that drive development, and agricultural markets that drive the phosphorus budgets of farms. The central question of applied limnology today is "How can we restore and sustain water quality, fisheries, and the other societal benefits of lakes in this new regime of regional to global control?" This question will not be answered by small-scale research, activism, or reactive management. It requires new, synthetic, fundamental studies of lakes and regulatory processes at multiple scales. The requisite understanding will come from a fusion of knowledge from natural and social sciences.

Concluding Remarks

Resilience can stabilize valuable ecosystems or undesirable ones. Both valuable and degraded states of lakes can be self-sustaining. Agriculture and urbanization switch lakes from valuable to degraded states by changing the processes that control water quality and fisheries. However, lake degradation is not a necessary consequence of agriculture and urbanization. High-quality lakes can be maintained in developed landscapes. Restoration of lakes to the valuable state requires interventions that shift economic and ecological controls at several scales. Sustainable restoration will link the economic benefits of clean lakes to incentives for conservation.

Acknowledgments

This chapter was inspired by the Resilience Network meeting on Little St. Simon's Island, April 1996. We thank Terry Chapin, Tony Ives, Mike Pace, Mary Power, Daniel Schindler, and an anonymous referee for helpful comments on the manuscript. We are grateful for support from the Pew Foundation, the National Science Foundation, and the National Center for Ecological Analysis and Synthesis.

Literature Cited

Adler, R., and J. C. Landman. 1993. *The Clean Water Act twenty years later*. Washington, D.C.: Island Press.

Arrow, K., B. Bolin, R. Costanza, P. Dasgupta, C. Folke, C. S. Holling, B.-O. Jansson, S. Levin, K.-G. Mäler, C. Perrings, and D. Pimentel. 1995. Economic growth, carrying capacity, and the environment. *Science* 268:520–521.

Blindow, I., G. Andersson, A. Hargeby, and S. Johansson. 1993. Long-term pattern of alternative stable states in two shallow eutrophic lakes. *Freshwater Biology* 30:159–167.

Campbell, D. G. 1996. Splendid isolation in Thingvallavatn. *Natural History* (June):48–55.

Caraco, N. F. 1993. Disturbance of the phosphorus cycle: A case of indirect effects of human activity. *Trends in Ecology and Evolution* 8:51–54.

Carpenter, S. R. 1988. Transmission of variance through lake food webs. Pp. 119–138 in *Complex interactions in lake communities*, edited by S. R. Carpenter. New York: Springer-Verlag.

Carpenter, S. R., D. Bolgrien, R. C. Lathrop, C. A. Stow, T. Reed, and M. A. Wilson. 1998. Ecological and economic analysis of lake eutrophication by nonpoint pollution. *Australian Journal of Ecology* 23:68–79.

Carpenter, S. R., T. M. Frost, J. F. Kitchell, T. K. Kratz, D. W. Schindler, J. Shearer, W. G. Sprules, M. J. Vanni, and A. P. Zimmerman. 1991. Patterns of primary production and herbivory in twenty-five North American lake ecosystems. Pp. 67–96 in *Comparative analyses of ecosystems: Patterns, mechanisms, and theories*, edited by J. Cole, S. Findlay, and G. Lovett. New York: Springer-Verlag.

Carpenter, S. R., and J. F. Kitchell, eds. 1993. *The trophic cascade in lakes*. Cambridge: Cambridge University Press.

Carpenter, S. R., J. F. Kitchell, K. L. Cottingham, D. E. Schindler, D. L. Christensen, D. M. Post, and N. Voichick. 1996. Chlorophyll variability, nutrient input, and grazing: Evidence from whole-lake experiments. *Ecology* 77:725–735.

Carpenter, S. R., C. E. Kraft, R. Wright, X. He, P. A. Soranno, and J. R. Hodgson. 1992. Resilience and resistance of a lake phosphorus cycle before and after food web manipulation. *Amerian Naturalist* 140:781–798.

Carpenter, S. R., and P. R. Leavitt. 1991. Temporal variation in a paleolimnological record arising from a trophic cascade. *Ecology* 72:277–285.

Carpenter, S. R., and M. L. Pace. 1997. Dystrophy and eutrophy in lake ecosystems: Implications of fluctuating inputs. *Oikos* 78:3–14.

Christensen, D. L., B. R. Herwig, D. E. Schindler, and S. R. Carpenter. 1996. Impacts of lakeshore residential development on coarse woody debris in north temperate lakes. *Ecological Applications* 6:1143–1149.

Christie, W. J. 1974. Changes in the fish species compositions of the Great Lakes. *Journal of the Fisheries Research Board of Canada* 31:827–854.

Cole, J. J., N. F. Caraco, G. W. Kling, and T. K. Kratz. 1994. Carbon dioxide supersaturation in the surface waters of lakes. *Science* 265:1568–1570.

Cornett, R. J., and F. H. Rigler. 1979. Hypolimnetic oxygen deficits: Their prediction and interpretation. *Science* 205:580–581.

Cottingham, K. L. 1996. Phytoplankton responses to whole-lake manipulations of nutrients and food webs. Ph.D. diss. University of Wisconsin.

Cottingham, K. L., and S. R. Carpenter. 1994. Predictive indices of ecosystem resilience in models of north temperate lakes. *Ecologist* 75:2127–2138.

Daniel, T. C., A. N. Sharpley, D. R. Edwards, R. Wedepohl, and J. L. Lemunyon. 1994. Minimizing surface water eutrophication from agriculture by phosphorus management. *Journal of Soil and Water Conservation* 49:30–38.

DeAngelis, D. L. 1992. *Dynamics of nutrient cycling and food webs.* New York: Chapman and Hall.

DeAngelis, D. L., S. M. Bartell, and A. L. Brenkert. 1989. Effects of nutrient recycling and food chain length on resilience. *American Naturalist* 134:778–805.

Driscoll, C. T., C. Yan, C. L. Schofield, R. Munson, and J. Holsapple. 1994. The mercury cycle and fish in Adirondack lakes. *Environmental Science and Technology* 28:137–143.

Edmondson, W. T. 1969. Eutrophication in North America. Pp. 124–149 in *Eutrophication: Causes, consequences, correctives.* Washington, D.C.: National Academy Press.

Frost, T. M., S. R. Carpenter, A. R. Ives, and T. K. Kratz. 1995. Species compensation and complementarity in ecosystem function. Pp. 224–239 in *Linking species and ecosystems,* edited by C. Jones and J. Lawton. London: Chapman and Hall.

Gunderson, L. H., C. S. Holling, and S. S. Light. 1995. *Barriers and bridges to the renewal of ecosystems and institutions.* New York: Columbia University Press.

Harper, D. 1992. *Eutrophication of freshwaters.* London: Chapman and Hall.

Hasler, A. D. 1947. Eutrophication of lakes by domestic drainage. *Ecology* 28:383–395.

Heck, K. L., and L. B. Crowder. 1991. Habitat structure and predator-prey interactions in vegetated aquatic systems. Pp. 281–299 in *Habitat structure: The physical arrangement of objects in space,* edited by S. S. Ball, E. D. McCoy, and H. R. Mushinsky. London: Chapman and Hall.

Hemond, H. F. 1990. Wetlands as the source of dissolved organic carbon to surface waters. Pp. 301–313 in *Organic acids in aquatic ecosystems,* edited by E. M. Perdue and E. T. Gjessing. New York: John Wiley and Sons.

Holling, C. S. 1973. Resilience and stability of ecological systems. *Annual Review of Ecology and Systematics* 4:1–23.

———. 1992. Cross-scale morphology, geometry, and dynamics of ecosystems. *Ecological Monographs* 62:447–502.

Holling, C. S., G. Peterson, P. Marples, J. Sendzimir, K. Redford, L. Gunderson, and D. Lambert. 1996. Self-organization in ecosystems: Lumpy geometries, periodicities and morphologies. Pp. 346–384 in *Global change and terrestrial ecosystems,* edited by B. H. Walker and W. L. Steffen. Cambridge: Cambridge University Press.

Howarth, R. W. 1991. Comparative responses of aquatic ecosystems to toxic chemical stress. Pp. 169–195 in *Comparative analyses of ecosystems,* edited by J. Cole, G. Lovett, and S. Findlay. New York: Springer-Verlag

Ives, A. R. 1995. Measuring resilience in stochastic systems. *Ecological Monographs* 65:217–233.

Jeppesen, E., J. P. Jensen, P. Kristensen, M. Sondergaard, E. Mortensen, O. Sortkjaer, and K. Olrik. 1990. Fish manipulation as a lake restoration tool in shallow, eutrophic, temperate lakes. 2: Threshold levels, long-term stability, and conclusions. *Hydrobiologia* 200/201:219–227.

Johnston, C. A. 1991. Sediment and nutrient retention by freshwater wetlands: Effects on surface water quality. *Critical Reviews in Environmental Control* 21:491–566.

Jones, C., and J. Lawton, eds. 1995. *Linking species and ecosystems*. London: Chapman and Hall.

Jones, R. I. 1992. The influence of humic substances on lacustrine planktonic food chains. *Hydrobiologia* 229:73–91.

Kitchell, J. F., ed. 1992. *Food web management: A case study of Lake Mendota*. New York: Springer-Verlag.

Kitchell, J. F., and S. R. Carpenter. 1993. Variability in lake ecosystems: Complex responses by the apical predator. Pp. 111–124 in *Humans as components of ecosystems*, edited by M. McDonnell and S. Pickett. New York: Springer-Verlag.

Kitchell, J. F., E. A. Eby, X. He, D. E. Schindler, and R. M. Wright. 1994. Predator-prey dynamics in an ecosystem context. *Journal of Fish Biology* 45:1–18.

Leopold, A. 1991. Land pathology. Pp. 212–217 in *The river of the mother of god and other essays by Aldo Leopold*, edited by S. L. Flader and J. B. Callicott. Madison: University of Wisconsin Press.

Levin, S. A., ed. 1996. Forum: Economic growth and environmental quality. *Ecological Applications* 6:12–32.

Likens, G. E. 1984. Beyond the shoreline: A watershed-ecosystem approach. *Internationale Vereinigung für Theoretische und Angewandte Limnology* 22:1–22.

———. 1992. *The ecosystem approach: Its use and abuse*. Oldendorf/Luhe, Germany: Ecology Institute.

Lodge, D. M. 1993. Biological invasions: Lessons for ecology. *Trends in Ecology and Evolution* 8:133–137.

Ludwig, D., B. Walker, and C. S. Holling. 1997. Sustainability, stability, and resilience. *Conservation Ecology* [online] 1:1–8 http://www.consecol.org/vol1/iss1/art7.

Maser, C., and J. R. Sedell. 1994. *From the forest to the sea: The ecology of wood in streams, rivers, estuaries, and oceans*. Delray Beach, Fla.: St. Lucie Press.

May, R. M. 1973. *Stability and complexity in model ecosystems*, 2nd ed. Princeton, N.J.: Princeton University Press.

McKnight, D., D. F. Brakke, and P. J. Mulholland, eds. 1996. Freshwater ecosystems and climate change in North America. *Limnology and Oceanography* 416:815–1149.

Moss, B. 1995. The microwaterscape: A four-dimensional view of interactions among water chemistry, phytoplankton, periphyton, macrophytes, animals, and ourselves. *Water Science and Technology* 3:105–116.

Moss, B., J. Stansfield, K. Irvine, M. Perrow, and G. Phillips. 1996. Progressive restoration of a shallow lake: A twelve-year experiment in isolation, sediment removal, and biomanipulation. *Journal of Applied Ecology* 33:71–86.

Naiman, R. J., J. J. Magnuson, D. M. McKnight, and J. A. Stanford. 1995. *The freshwater imperative*. Washington, D.C.: Island Press.

National Research Council. 1992. *Restoration of aquatic ecosystems: Science, technology, and public policy*. Washington, D.C.: National Academy Press.

———. 1993. *Soil and water quality: An agenda for agriculture*. Washington, D.C.: National Academy Press.

O'Neill, R. V., D. L. DeAngelis, J. B. Waide, and T. F. H. Allen. 1986. *A hierarchical concept of ecosystems*. Princeton, N.J.: Princeton University Press.

Osborne, L. L., and D. A. Kovacic. 1993. Riparian vegetated buffer strips in water quality restoration and stream management. *Freshwater Biology* 29:243–258.

Pimm, S. L. 1984. The complexity and stability of ecosystems. *Nature* 307:321–326.

Pimm, S. L., and J. H. Lawton. 1977. Number of trophic levels in ecological communities. *Nature* 268:329–331.

Porcella, D. B., J. W. Huckabee, and B. Wheatley, eds. 1995. *Mercury as a global pollutant*. Dordrecht, Netherlands: Kluwer Academic Publishers.

Post, D. M., S. R. Carpenter, D. L. Christensen, K. L. Cottingham, J. R. Hodgson, J. F. Kitchell, and D. E. Schindler. 1997. Seasonal effects of variable recruitment of a dominant piscivore on food web structure. *Limnology and Oceanography* 424:722–729.

Postel, S., and S. R. Carpenter. 1997. Freshwater ecosystem services. Pp. 195–214 in *Nature's services*, edited by G. Daily. Washington, D.C: Island Press.

Power, M. E., D. Tilman, J. A. Estes, B. A. Menge, W. J. Bond, L. S. Mills, G. Daily, J. C. Castilla, J. Lubchenco, and R. T. Paine. 1996. Challenges in the quest for keystones. *BioScience* 46:609–620.

Rasmussen, J. B., R. B. Rowan, D. R. S. Lean, and J. H. Carey. 1990. Food chain structure in Ontario lakes determines PCB levels in lake trout *Salvelinus namaycush* and other pelagic fish. *Canadian Journal of Fisheries and Aquatic Science* 47:2030–2038.

Rudd, J. W. M., C. A. Kelly, D. W. Schindler, and M. A. Turner. 1988. Disruption of the nitrogen cycle in acidified lakes. *Science* 240:1515–1518.

Rudstam, L. G., R. C. Lathrop, and S. R. Carpenter. 1993. The rise and fall of a dominant planktivore: Direct and indirect effects on zooplankton. *Ecology* 74:303–319.

Scheffer, M. 1991. Fish and nutrients interplay determines algal biomass: A minimal model. *Oikos* 62:271–282.

Scheffer, M., S. H. Hosper, M.-L. Meijer, B. Moss, and E. Jeppesen. 1993. Alternative equilibria in shallow lakes. *Trends in Ecology and Evolution* 8:275–279.

Schindler, D. W. 1977. The evolution of phosphorus limitation in lakes: Natural mechanisms compensate for deficiencies of nitrogen and carbon in eutrophied lakes. *Science* 195:260–262.

———. 1990. Experimental perturbations of whole lakes as tests of hypotheses concerning ecosystem structure and function. *Oikos* 57:25–41.

Schindler, D. W., S. R. Carpenter, K. L. Cottingham, X. He, J. R. Hodgson, J. F. Kitchell, and P. A. Soranno. 1996a. Food web structure and littoral coupling to pelagic trophic cascades. Pp. 96–108 in *Food webs: Integration of pattern and dynamics*, edited by G. A. Polis and K. O. Winemiller. New York: Chapman and Hall.

Schindler, D. W., P. J. Curtis, B. R. Parker, and M. P. Stainton. 1996b. Consequences of climate warming and lake acidification for UV-B penetration in North American boreal lakes. *Nature* 379:705–708.

Schindler, D. W., T. M Frost, K. H. Mills, P. S. S. Chang, I. J. Davies, D. Findlay, D. F. Malley, J. A. Shearer, M. A. Turner, P. J. Garrison, C. J. Watras, K. Webster, J. M. Gunn, P. L. Brezonik, and W. A. Swenson. 1991. Comparisons between experimentally and atmospherically acidified lakes. *Proceedings of the Royal Society of Edinburgh* 97B:193–226.

Soltero, R. A., L. R. Singleton, and C. R. Patmont. 1992. The changing Spokane River watershed: Actions to improve and maintain water quality. Pp. 458–478 in *Watershed management*, edited by R. J. Naiman. New York: Springer-Verlag.

Stow, C. A., S. R. Carpenter, C. P. Madenjian, L. A. Eby, and L. J. Jackson. 1995. Fisheries management to reduce contaminant consumption. *BioScience* 46:752–758.

Tilman, D. 1996. Biodiversity: Population versus ecosystem stability. *Ecology* 77:350–363.

van Donk, E., and R. D. Gulati. 1995. Transition of a lake to turbid state six years after biomanipulation: Mechanisms and pathways. *Water Science and Technology* 32:197–206.

Vollenweider, R. A. 1976. Advances in defining critical loading levels for phosphorus in lake eutrophication. *Memorie dell'Istituto Italiano di Idrobiologia* 33:53–83.

Walker, B. H. 1992. Biological diversity and ecological redundancy. *Conservation Biology* 6:18–23.

Wetzel, R. G. 1983. *Limnology*. Philadelphia, Penn: W. B. Saunders.

———. 1990. Land-water interfaces: Metabolic and limnological regulators. *Internationale Vereinigung für Theoretische und Angewandte Limnology* 24:6–24.

———. 1992. Gradient-dominated ecosystems: Sources and regulatory functions of dissolved organic matter in freshwater ecosystems. *Hydrobiologia* 229:181–198.

4

The Baltic Sea: Reversibly Unstable or Irreversibly Stable?

Bengt-Owe Jansson and AnnMari Jansson

The Baltic Sea with its drainage basin exemplifies a large-scale, dynamic and highly productive system shaped by natural processes as well as human activities. Its nested hierarchy spans a multinational, highly industrialized and exploited watershed, an enclosed brackish-water sea with several distinct subsystems and an intricate web of bacterial processes at the microscale. The resilience of this huge system is continuously tested by local and regional as well as global forces.

History

Since the last glaciation, the Baltic Sea has undergone a series of varying freshwater and marine stages as the connection to the North Sea has alternatively narrowed and widened. Fishing and hunting for seal and waterfowl constituted the traditional base of subsistence for people living on the coast of the Baltic Sea. For centuries, the rise and fall of the herring stock was critical to the well being of the coastal communities.

The pristine nature of the sea has successively changed due to the increasing population pressure in the catchment area (figure 4.1). Land ownership and land use practices show a pulse-wise evolution forced by immigrations of various tribes and alternately dominated by religious and political powers. Innovations in shipbuilding and navigation stimulated the development of commercial centers, some of which persisted and grew in size to become what are now present-day cities. Distinct flips in the exploitation pattern of forests, freshwater systems, minerals, and other natural resources have occurred due to major

Figure 4.1. The Baltic Sea and its drainage area that is four times larger in size than the sea and incorporates nine countries. The Baltic Proper is permanently stratified with a low salinity layer at the surface and a more saline, often anoxic bottom layer. The other basins are well ventilated. Some 90 million people who reside in the drainage area have successively pushed the Baltic Proper toward the border of its stability domain.

inventions, such as the waterwheel and steam engine, and through the use of fossil fuel and electricity.

In the nineteenth and twentieth centuries, the human population increased about four times and the use of fossil energy about thirty times (Kaasik 1989). Now, the activities of some 90 million people is the overriding force in the transformation of the Baltic, as it is at every hierarchical level of the global system (Odum 1971).

Recent Conditions of the Baltic Sea Ecosystem

The current conditions of the Baltic Sea are clearly shaped by the dynamics of the higher system (figure 4.2, circles). Increasing inputs of nitrogen and phosphorus from land run-off (agriculture, forestry) and atmosphere (traffic, industry) in the twentieth century have increased the primary production of plankton algae by 30–70 percent, the zooplankton production by 25 percent, and the

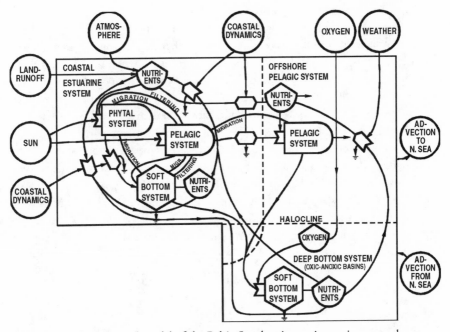

Figure 4.2. A conceptual model of the Baltic Sea showing major environmental inputs, subsystems, and couplings. The primary halocline is a natural border between the shallow systems and the deep basin systems. The functional border between the coastal-estuarine region and the offshore region is situated in the area of upwelling/downwelling and coastal jets (Jansson 1981).

sedimentation of organic carbon by 70–190 percent (Elmgren 1989). Major consequences of this increased loading are a decreased transparency of the off-shore waters (a reduction of 2.5–3.0 meters since 1914, Sandén and Håkansson 1996) and extended periods of anoxia in deep waters (Matthäus and Schinke 1994). High concentrations of PCBs and DDT in top carnivores such as the common seal, alks, and the white-tailed eagle and high concentrations of mercury in alks, ospreys, and pike (Jensen et al.1969; Hook and Johnels 1972) appeared in the 1960s as alarming evidence of the detrimental impacts of industrialized society. These impacts occurred while fish harvesting increased ten-fold (since the 1950s), causing significant shifts of species in the marine food chain. Today, the stocks of herring as well as sprat and cod are influenced largely by fishing intensity and by fluctuating prices on the fish market. Constituting only about one-thousandth of the total ocean area, the Baltic Sea has at its best delivered close to 1 percent of the total world catch.

The Future Baltic?

Against this background, what can we expect of the future development of the Baltic Sea? How resistant is it to changes in the driving forces? Comparative studies of various complex systems have shown that they usually have several stability domains, and that drastic, persistent changes in the boundary conditions (variables at the next higher system level) may cause them to flip to another domain with different properties (Holling et al. 1994). Are these dynamics applicable to the Baltic Sea—the marine part of the Baltic Basin? Can we talk of a resilient sea without considering the dynamics of its major subsystems?

In the following, we shall focus on the dynamics of the marine ecosystem to evaluate its capacity to withstand further nutrient loading and future surprises. To do so we must look for the degree of nestedness of the system, the scales of change and patterns in historical time, and the key variables responsible for maintaining its stability. Emphasis is put on the role of nutrients, while the problems of toxic substances are more superficially treated, since they are given substantial attention in other writings about the Baltic Sea. The last section is devoted to coevolution of natural and human systems and the couplings between them.

The Baltic Ecosystem

The Baltic Sea is a single large-scale system with emergent properties, as described in this section. It is composed of three main subsystems whose interactions govern many of the dynamics of the whole.

Characteristics of the Sea

The Baltic Sea is a large sea, nearly 400,000 square kilometers, and is geo-
morphologically divided into three main basins (from south to north): the
Baltic Proper, the Bothnian Sea, the Bothnian Bay (figure 4.1). The coastline
extends through rocky archipelagos along the Swedish and Finnish coasts,
exposing large hard-bottom areas to a luxuriant growth of macroscopic algae.
The southern coast of the eastern Baltic countries (Poland, Germany, and
Denmark) is low and sandy. The Gulf of Finland and the Gulf of Riga are
large bays, the former more open to the Baltic Proper than the latter. The
Baltic Proper, the main focus of this chapter, is separated from the northern
basins by the shallow Archipelago Sea and from the North Sea by the shallow
straits between Denmark, Germany, and Sweden. The large volume contributes to a residence time of the Baltic water of twenty-five years (table 4.1).
A permanent salinity stratification is caused by freshwater run-off from the
drainage area, which induces an inflow of heavy saline North Sea water
through the shallow and narrow straits. In addition to the regular, annual,
exchange of water, intermittent pulses of North Sea water are pushed into the
Baltic caused by high- and low-pressure systems over the larger North Atlantic
system. The stream of Atlantic water flows counterclockwise around the island
of Gotland and joins the Swedish southbound coastal current flowing out
through the straits. This large-scale circulation establishes the stable salinity
gradient and thereby the borders between the three major subsystems. The circulation pattern maintains a surprisingly stable surface salinity gradient from

Table 4.1. Oceanographic data of the
Baltic Sea

Description	Measure
Area, total	415.000 km^2
Area, Baltic Proper	267,000 km^2
Length (N-S)	1,300 km
Width (W-E)	1,200 km
Average depth	60 m
Maximum depth	459 m
Sill depth	17 m
Volume	21,700 km^2
Residence time of water	25 years
Tides	2–3 cm
Total drainage area	1,641,650 km^2

10 parts per thousand in the south to practically freshwater in the inner part of the Bothnian Bay. The slow water exchange makes the Baltic an effective sediment trap but also makes it prone to accumulation of nutrients and toxic elements. The Baltic is also a cold sea, the northern part covered by fast ice during seven to eight months of the year and the bottom water holding a temperature of about 4°C year-round.

The low salinity has limited the oceanic immigration of salinity-sensitive organisms to a low-diversity community of plants and animals. There are few species in each functional group or guild (for example, primary producers, nitrogen fixers, filterfeeders, carnivores, decomposers). Of the 1,500 macroscopic animals found off the coast of southern Norway, only some 150 occur inside the sills, and in the northern Baltic Proper the number of marine animals decreases to around fifty. These species still maintain large populations of fertile individuals but are mostly of small body size due to slow growth.

Subsystems

Three major systems can be distinguished in the Baltic Sea: the coastal, the offshore pelagic, and the deep bottoms (figure 4.2). The functional border between the coastal and pelagic subsystem is defined where upwelling and downwelling events with accompanying coastal jet currents operate (Jansson et al. 1984), which in the Baltic Proper often coincides with the 50-meter isobath. The boundary of the deep bottom subsystem is the primary halocline, the steep salinity gradient between the surface and the bottom water at a depth of 60–70 meters.

COASTAL SUBSYSTEM

The coastal system is an ecotone utilizing resources both from land and from the offshore area (Jansson 1980). It is functionally crucial for the total Baltic system as a sediment trap and buffer for the flows of nutrients and toxic substances from land toward the offshore areas, as a spawning and nursery ground for economically important fish species, and as a recreational area for an increasing number of people in Baltic Europe.

From a global perspective the coastal system may be considered as a "selfish" system characterized by tight internal cycling and a continuous import of, for example, nutrients coming from land and in even larger amounts from offshore. There is practically no tide in the Baltic, which means a restricted circulation of sheltered areas. The turnover time of the water in the coastal subsystem has been estimated at around one month (Hinrichsen and Wulff 1998).

The coastal-estuarine zone is the most diversified subsystem, both in

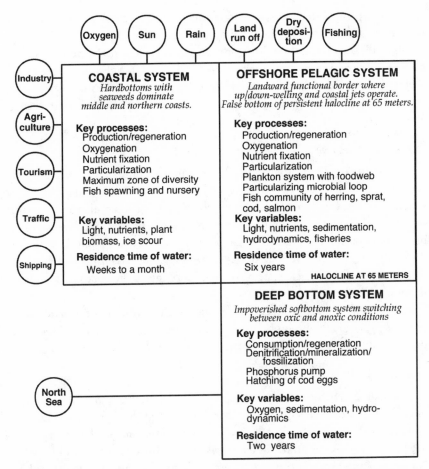

Figure 4.3. Structural and functional characteristics of three subsystems in the Baltic Sea. The high-diverse coastal system has locally severe problems of eutrophication and hazardous wastes. The offshore pelagic system exports its surplus of organic material, and only a summer thermocline may cause an intermittent shortage of critical material. The deep bottoms are the ultimate sink for particulate organic material and phosphorus, and the site for large-scale denitrification. They constitute the alarm clock for stagnation of bottom water and impoverished conditions for the cod fisheries.

species and habitats, performing an array of key functions (figure 4.3). Solar energy fixation is high due to abundant light and turbulent water conditions. This promotes oxygenation, nutrient fixation, and particularization of dissolved matter to the benefit of numerous grazers such as crustaceans, snails, and filter-feeders like clams and mussels. Herbivores feeding on macroalgae and higher plants are rare, and decomposition and regeneration of nutrients from this source of organic material occurs mainly through detritus eaten by soft-bottom organisms. The fish community is diverse, with different assemblies in the clear waters of the rocky shores and in the turbid water of the more shallow bays bordered with reed and pond weed. The key variables of the coastal subsystem are light, nutrients, plant biomass, and the ice scour during winter time.

OFFSHORE PELAGIC SYSTEM

According to our classification, the offshore subsystem does not include a typical sediment phase. The persistent halocline acts as a false bottom, where the sinking leftovers from the pelagic system slow down and accumulate in the sharp salinity gradient. Here decomposition in a diverse microbial food web causes high concentration of nutrients in the pelagic water. It takes about six years for the total volume of surface water to be exchanged (Hinrichsen and Wulff 1998). Key variables of the pelagic system are light, nutrients, sedimentation, hydrodynamics, and fisheries.

The annual primary production of about 150 grams carbon per square meter supports a pelagic foodweb of copepods, cladocerans, mysids, and fish populations, including economically important herring, sprat, cod, and salmon.

DEEP BOTTOM SUBSYSTEM

The oxic-anoxic basins cover the deepest parts of the Baltic Sea, the primary basins in the Baltic Proper being the Bornholm Deep, the Western and Eastern Gotland Basins, and the Gdansk Deep. The turnover time of the water of the deep bottom system is on the order of two years (Hinrichsen and Wulff 1998). No primary production exists, due to lack of light. The system is dominated by consumption and decomposition of organic matter part of which is remineralized inorganic nutrients prone to transportation to the other systems. Another part, out of reach of bioturbation by organisms, is successively buried in the bottom sediments and slowly fossilized. Denitrification by bacteria is an important property of the deep bottom system (Shaffer and Rönner 1984).

The organism assembly is an impoverished soft-bottom community dominated by *Macoma balthica*, the Baltic clam, with a few hydrogen sulfide resistant bristleworms, crustaceans, and nematodes (Segerstråle 1957). Most fish avoid waters with an oxygen content of less than 2 milliliters per liter, which makes the deeps used only sparingly as foraging areas. On the other hand, the deep waters serve an important role in the recruitment of cod larvae due to the higher salinity, which is necessary for eggs to develop floating in the water.

SUBSYSTEM COUPLINGS

As can be seen from the above descriptions, the three main subsystems have different functions in the total system of the Baltic Sea. There is, however, an important pulsing exchange between them forced both by the weather system of the North Atlantic and by the genetic programming of migrating fish and other organisms. The effectively photosynthesizing set of fast and slow primary producers in the nearshore waters form a loaded "pantry" for consumers many of which live as adults in the offshore pelagic system and migrate to the coast to spawn in spring and summer when there is abundant food and protection for the young fry. The seabirds in the coastal zone operate a feedback flow of material from sea to land and exert considerable bioturbation locally in the shallow bays, where numerous schools of eiderducks, goldeneyes, and tufted ducks forage on the bottoms. Forced by increased turbulence during periods of heavy wind, some of the settled surplus of particulate matter is successively transported along the bottom slopes toward the deep bottoms for breakdown or fossilization. A feedback loop of regenerated inorganic nutrients reaches the pelagic offshore system from below during weather-related upwelling events, stimulating plankton blooms and subsequent sedimentation.

Of overriding importance for the exchange between the subsystems are the intermittent inflows from the North Sea. The inflowing water first penetrates into the deep basins in the south, but the salt and nutrients are gradually mixed into the offshore pelagic system during the northward course of the bottom current.

Wind is a major force for the water transports. Persistent southwesterly winds cause upwelling of deeper, nutrient-rich water along the eastern coasts, sometimes inducing plankton blooms (Jansson 1978; Karhu et al. 1995). Similarly, northeasterly winds induce downwelling, increasing the downward transportation of organic matter. The coupled alongshore coastal jets transport considerable amounts of water, estimated to be as large as one-tenth of the Gulf Stream (Shaffer 1975). Exchange between the three subsystems thus exhibits a pulse-like behavior.

Scales of Change in the Baltic: Cycles, Trends, and Flips

Ecosystem dynamics in the Baltic are complex and include cyclic patterns, gradual changes, and sudden discontinuities. Each type of change occurs not just at a single scale, but at a wide range of scales in space and time.

Cyclic Patterns

A strong solar pulse creates a pronounced annual pattern in the Baltic Sea typical of ecosystems at these high latitudes. During summer the primary production is intensive during the day, and the shallow water is often supersaturated with oxygen. Many fish are day-active, roaming in the sunlit waters, which are seemingly barren of other macroscopic animals. During the night, the lower life among the seaweeds swarms in the now-dark surface waters (Jansson and Källander 1968). The winter period, when the plants are living on stored organic matter in the absence of absorbing photosynthesis, is dominated by decomposition, and nutrients accumulate in the bottom waters. Many animals from the shallow waters move to deeper areas during wintertime, whereas some cold-water species, for example, members of the sculpin family with arctic origin, migrate to the cold shallow waters to spawn at a few meters depth (Jansson 1980).

Upwelling of bottom water in early spring and large input through river run-off, especially after winters with lots of snow and rain (figure 4.4a), create a peak of nutrients in the surface waters. A spring bloom of phytoplankton starts in March, when nearly half of the synthesized organic matter settles to the bottom to provide about half of the annual food needs of the soft-bottom community (Elmgren 1978). A summer pulse of pelagic green algae leaves mostly low-calorie fecal pellets for the soft-bottom animals. A conspicuous bloom of cyanobacteria (bluegreens) in July–August is followed by a final pulse of dinoflagellates and diatoms before respiration-decomposition processes again take over in the winter darkness.

The hard-bottom algae and soft-bottom plants follow a similar pattern, with an early spring pulse of sessile diatoms on the rocky bottoms, now cleaned from macroalgae and loose sediments by ice scour and winter storms. A green and brown longshore border of annual, filamentous green algae spreads, and brown veils of filamentous annuals soon cover most hard substrates, including the surfaces of perennial bladderwrack (*Fucus vesiculosus*).

Many fish populations utilize these alternating bursts of organic matter production. Herring, which accumulate fat reserves when foraging on large zooplankton in late summer, spend winters in the cold coastal waters maturing for the spawning in May. The fry have rich food in the rapidly warming shallow waters and migrate offshore after a few months. Pike and perch deposit their strings of eggs in the shallow vegetation, a habitat and source of food not only for new generations of fish but also for many other members of the coastal food web.

The pulsed production in the surface layers of the sea generates corresponding bursts in consumer/degradation processes at the higher trophic levels, but some organic matter is also transferred downward to the pelagic stratified layers and the deeper soft bottoms.

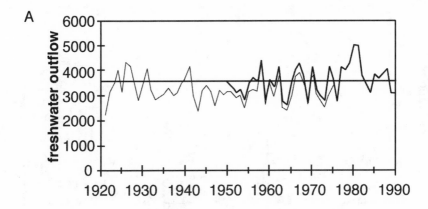

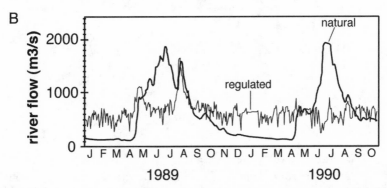

Figure 4.4. (A) Annual variability of freshwater flow to the Baltic Proper (Bergström and Carlsson 1994). 1981–1990 was the wettest in seventy years. (B) Effect of hydropower development on flows in one of the bigger Nordic rivers, Luleälven. On top of the monthly mean changes there are day-to-day fluctuations in the efforts to meet the power demand (Bergström and Carlsson 1994).

On top of these annual cycles are the pulses of North Sea water caused by the distribution of high and low pressures over the Atlantic (figure 4.5). These seem to come in clusters rather than being cyclic (Matthäus and Schinke 1994). A ten-year absence of major inflows between 1983 and 1993 has been explained by an increased zonal circulation linked to intensified precipitation over the Baltic Drainage Basin, with subsequent increased river run-off. Variations in the latter (figure 4.4b) seem to have greater impact than originally thought (Schinke and Matthäus 1998).

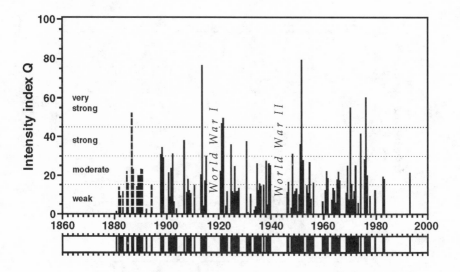

Figure 4.5. Major inflows of highly saline and oxygenated water from the North Sea to the Baltic Proper between 1880 and 1994. The pattern below is generalized inflow clusters (Matthäus 1995). Observe the low frequency of inflows after 1980.

Trends

In the last century, major systems variables of the Baltic Sea have changed. Records of freshwater run-off show increasing fluctuations between years since the 1920s, with the highest peak in 1981 (figure 4.4b). The salinity of the deep water in the central Baltic has varied with the inflows of saline water from the North Sea, but after the mid-1970s a downward trend is apparent (Matthäus 1995). The Bornholm Basin, which acts as a kind of buffer against the inflows, has undergone a period of oceanization from the 1950s through the 1970s, when an increased frequency of marine animal species was observed in the biological community. A previous dominance of arctic fauna was broken by the stagnation periods during the 1950s, and in the last decades cosmopolitan species have become more common. Cold-adapted suspension feeders were replaced by hardy, nonselective deposit-feeders (Leppäkoski 1975). The appearance of the marine jellyfish *Cyanea capillata* in the shallow areas of the Northern Baltic Proper was another visible sign of the rise in salinity in the 1970s (Jansson 1978).

There is a negative correlation between the strength of the freshwater run-off and the periods of saltwater inflows—the larger the run-off, the lower the inflows of saline water (Matthäus 1995). A long stagnation period after the mid-

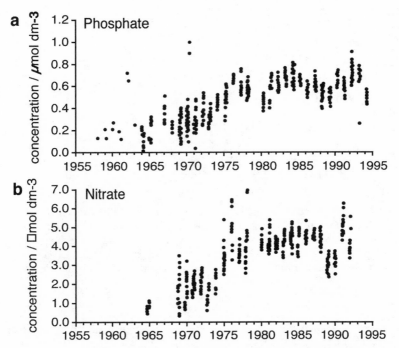

Figure 4.6. Long-term variation of inorganic nutrient concentrations found in the winter surface layer of the Eastern Gotland Basin (Matthäus 1995). Winter values reveal the comparable size of the nutrient pools, later masked by the uptake by the light-stimulated phytoplankton.

1980s and high precipitation during the 1980s caused a decrease in salinity in the Northern Baltic Proper. The subsequent decrease in some major marine zooplankton (Vuorinen et al.1998) had a negative effect on the growth of herring (Flinkman et al. 1998).

Nutrients show an increasing trend during the 1900s (figure 4.6), a result of eight- and four-fold increases in the loads of phosphorus and nitrogen respectively (Larsson et al. 1985). After the middle of the 1960s, the winter concentrations in the surface layer of the open Baltic showed a strong increase, but these leveled off during the 1980s (Matthäus 1995).

Since the early 1900s, the transparency of the water in the Baltic Proper has decreased by approximately 2.5–3.0 meters (Sandén and Håkansson 1996). This reflects an almost 50 percent increase in pelagic primary production between the 1940s and the 1980s (Elmgren 1989). The reduction of submarine light has caused a decreased depth in the distribution of the perennial bladder-

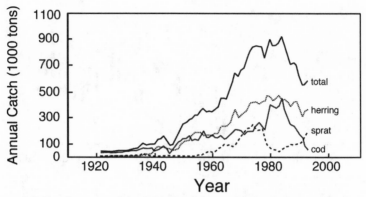

Figure 4.7. Development of the Baltic Sea fisheries during the 1900s. The large catches of cod in the late 1970s and early 1980s were due to successful spawning during the mid-1970s because of saltwater inflows. Overfishing generated the sharp decline of sprat landings in the late 1970s. Catch data before 1963 from Hansson and Rudstam (1990); later data from ICES Cooperative Research Report 210 (1995).

wrack since the 1940s (Kautsky et al. 1986; Wachenfeldt et al. 1986), a trend that also seems to have leveled off in the 1980s.

For the fish community, long-term quantitative data only exist in the form of fishery statistics on annual catches of main commercial species (figure 4.7). The steep increase in total catch, although primarily an effect of improved technical efficiency of the fishery, can also be explained by an increase in biomass of the soft-bottom macrofauna above the halocline (Cederwall and Elmgren 1980), which provide more abundant food for herring and sprat (Thurow 1989). Between 1970 and 1985, however, the standing stocks of those fish species decreased by 30 percent due to heavy fishing for the fishmeal industry (Thurow 1989). Since then sprat has become increasingly abundant. As mentioned before, cod needs good oxygen conditions below the halocline for successful hatching of its eggs; eutrophication consequently has a direct influence on the strength of the year classes of this fish species. Strong inflows of oxygen-rich saline water in the late 1970s created the optimum conditions for a successful spawning of the cod. This resulted in unprecedented large cod catches in the beginning of the 1980s. However, too-heavy fishing led to a fast decline of the cod stock, which has not recovered since then.

In summary, the dynamics of the Baltic 1900s can be classified into three main periods (table 4.2). These are:

Stage I, the period before 1940, is characterized by clear water, fairly low primary production, luxuriant perennial algal belts, and little filamentous algae

Table 4.2. Developmental stages in the Baltic Proper system during the twentieth century. Since 1980, with little increase in nutrient levels, annual and perennial seaweed cover has stabilized. Changes in fish community indicate more complex couplings of predation, overfishing, and periods of anoxia

Variable	Pre-1940	1940–1980	1980–2000
Nutrients	Low	Increasing	Leveling off
Secchi disc (m)	11	NA	8
Pelagic primary productivity[a]	79–103	134	160
Seaweeds (perennial)	Luxuriant	Decreasing	Leveling off
Seaweeds (annual)	Low	Increasing	Stable
Herring, sprat	Common	Decreasing	Increasing
Cod	Common	Increasing	Decreasing
Salmon (wild)	Common	Decreasing	Nearly extinct
Salmon (reared)	None	Increasing	Leveling off

[a]Elmgren 1989.

(e.g., bladderwrack almost free from epiphytes). The stocks of major commercial fish species herring, cod, flounder, and wild salmon were probably larger, although this does not show up in the landing statistics as the technical capacity to fish was low, due to small-sized fishing vessels, primitive equipment, and limited access to fossil fuel. This period represents the still-oligotrophic state of the Baltic Sea ecosystem.

Stage II represents the period 1940–1980, when nutrients were steadily increasing, with a decrease of the bladderwrack due to the explosion of brown filamentous algae. Increasing landings of herring and cod tell of good stocks. Oligotrophy has changed to eutrophy.

Stage III, the stagnation period after 1980, implies a leveling off of nutrients and signs of silica deficiency in some areas, high production of pelagic and filamentous algae mainly limited by light, and displacements in the fish community due to decreasing cod populations.

Anthropogenic Indicators

Toxic substances in the marine ecosystems show trends similar to those of the nutrients (figure 4.8). Concentrations of heavy metals such as lead, copper, mercury, zinc, and cadmium in the sediments increase after 1940s, reach a peak at 1980, and then begin to decrease (Thurow 1989; Bernes 1988). The scientific

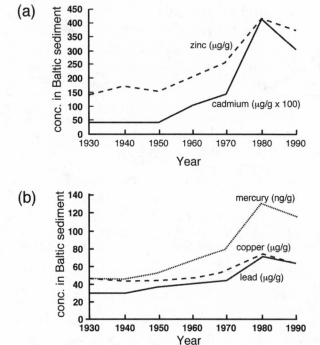

Figure 4.8. Time course of metals found in Baltic Proper sediments, deposited from 1930 to 1990 (redrawn from Jonsson 1992). *(a)* High-ranging concentrations (up to 450 µg/g) and *(b)* lower-level concentrations.

and public awareness of the health risks of these and other toxic pollutants from industrialized society began when alarmingly high levels of mercury were found in fish from both lakes and coastal waters in Scandinavia. This initiated an intense scientific activity focusing on the importance of the trophic structure of the ecosystem (Johnels et al.1967) and leading to a ban on mercury and changes in the chemical processes of the pulp industry.

Production and use of PCBs and DDT started in the early 1900s. Their concentrations in seals and birds increased rapidly after 1940 and reached a maximum in the middle of the 1970s. Threatened by extinction, the common seal evoked a vigorous public debate on the character of industrial society and still-expanding research on the occurrence and effects of chemical compounds in ecosystems (Jensen et al.1969; Olsson and Reutergård 1986). Now banned in the Baltic region, they have significantly decreased after 1980, with seal and bird populations recovering at a fast rate and the white-tailed eagle back in numbers of 1800 levels (BSEP 1996).

Main Switches and Flips in the System

During the first decades of World War II, the stress on the Baltic ecosystem is evidently increasing rapidly as result of the intensified human activities in the drainage area. The reactions of society to those changes constitute a feedback, affecting the economic system in the form of, for example, consumer aversion to Baltic fish products, demand for environmental education and information, and development of new institutions. The necessity of treating the Baltic Sea and its drainage area—including the human systems—as one big system becomes increasingly obvious. But to develop a flexible system for managing of the Baltic Sea one must find out if it has any discontinuities, which should induce special caution in human resource use.

There are several forcing functions that "test" the potential resilience of the Baltic ecosystem. Again, it is important to stress the large spatial and temporal scale of the Baltic and the heterogeneity of habitats. The same stressor may affect the separate subsystems in very different ways.

As emphasized in earlier paragraphs, a major driving force with profound effects on all subsystems is the nutrient load. As natural parts of the system, nitrogen and phosphorus influence most other flows in the system. Figure 4.9 is an effort to depict the negative (–) or positive (+) effects of one compartment on another. An increase in the total nutrient load stimulates both phyto- and zoo-plankton to higher production, increasing turbidity with negative effects for the perennial macrophytes, which are already stressed by the shading effects of increasing, annual, epiphytic filamentous algae. Increased turbidity has a negative impact on the clear-water fish in the coastal areas, including fish such as pike, perch, and whitefish, whereas the excessive production favors "scrap fish," such as bream, roach, and ide. The increased plankton production causes heavy sedimentation on the soft bottoms, increasing the oxygen consumption of the seabed, especially by the aerobic bacteria populations. In addition, toxic algal blooms are triggered by nitrogen:phosphorus ratios above the normal 7:1. Their toxicity makes them unattractive as food for pelagic herbivores. The increased plankton production favors the growth of the pelagic herring and sprat populations. This means more food for adult salmon and cod. On the other hand, the numerous large-sized sprat regulate the cod population by consuming the hatched cod larvae.

Not shown in the diagram is the possible decrease in quality of herring as food

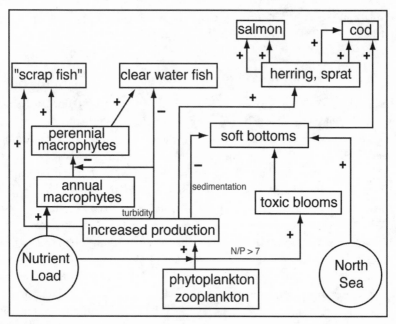

Figure 4.9. Conceptual diagram indicating the crucial role of nutrients in the Baltic ecosystem. A + means an increase of the target hit by the arrow; a – indicates a decrease. Increased production has a negative effect on perennial macrophytes through the increase of annual, filamentous macrophytes and through increased turbidity. This also affects clear-water fish, such as whitefish, perch, and pike. The resulting increased sedimentation, including decomposing toxic algal blooms, decreases the food resources of soft bottoms for bottom-living fish such as cod. Pelagic fish such as herring and sprat, benefit from increased pelagic production, which in turn provides more food for both salmon and cod.

for salmon, caused by a shift in the pelagic food web. The originally strong component of diatoms in the normal plankton blooms, which contains the important carotenoid astaxanthine, has decreased during the 1990s due to shortage of silica (Snoeijs 1999). This shortage is caused by the heavy diatoms themselves, which, stimulated by the increasing nutrients, are stripping the water column of silica through their fast sedimentation. Herring and sprat therefore transfer less of this crucial antioxidant to the salmon from their pelagic and benthic food items, the immune system of the salmon is weakened, and the so-called M-74 syndrome appears, causing a mortality of 80–90 percent of the salmon larvae (Bengtsson 1998). This intricate network of processes from the straightforward stimulation of primary production to the erosion of the immune defense of the salmon via the changing food web demonstrates the central role of nutrients and their balanced status in the Baltic Sea ecosystem (Karjalainen 1999).

We can recognize two types of flips: those that are intrinsic in the respective subsystem and those that involve several subsystems. To the former belongs the switch from dominant perennial and palmated seaweeds such as the bladder-wrack to filamentous annuals, a flip that appeared almost like an explosion during the early 1970s as a reaction to increased nitrogen levels. It affects not only hard bottoms but also shallow soft bottoms, where the stems of reed and pond weeds now wear "skirts" of brown *Pilayella* or green *Enteromorpha*. Changes in the fish community exemplify how migratory key organisms may influence the state of several subsystems, an example of the second type of flip. Herring, for example, spawn in the coastal subsystem and live as adults in the offshore pelagic. Thus the strength of offshore populations of herring is linked to the effects of eutrophication in the coastal subsystem. The extensive cover of fila-mentous algae on the bottoms of the spawning areas of herring increases the mortality of its eggs to 75 percent, compared to less than 10 percent in marine waters (Aneer 1987). Also, excess sedimentation in Baltic shallow waters reduces the area available for successful spawning.

The deep soft-bottom systems are more closed than those of the coastal and offshore areas. The intermittent pulses of North Sea water operate like switches not only for salt and oxygen but also for assemblies of organisms. A classical example is the inflows in 1923 when haddock larvae were brought into the southern Baltic, giving a ten-fold increase in haddock catches during 1925–1926, which then went down when the population was ready to spawn and returned to the North Sea. The increase in organic matter of the often anoxic bottoms has caused the switch from a previous suspension feeding com-munity of the *Macoma*-association, including several species of mussels and clams, crustaceans, and bristleworms to the low-diverse community of hydrogen sulfide-resistant deposit-feeders, dominated by the bristleworm *Harmothoe*.

The flipping between oxic and anoxic states in the deep bottom system seems to wear off over time (figure 4.5). The increasing sedimentation of organic matter at least up to the 1980s has resulted in increasing amounts of hydrogen sulfide accumulating during the anoxic state. This means that to oxidize deep bottom systems successfully increasing amounts of inflowing oxygen-rich water are needed, firstly to oxidize the hydrogen sulfide, secondly to allow recoloniza-tion by higher, aerobic organisms, and thirdly to sustain the established popula-tions. Hence the deep bottom subsystem has changed into a permanently oxygen-free state, devoid of higher life.

At the scale of the whole Baltic, the transfer from an oligotrophic state to a eutrophic one during 1940–1980 can be recognized as a large-scale flip, taking place over a period of forty years.

Resilience in the Baltic

In order to understand the principal behavior of the large, complex Baltic ecosystem, we have tried to fit it into the resilience concept developed by Holling (1976). Holling argues that the main spatial and temporal patterns of complex systems can be traced to a small set of variables operating at different scales in time and space. The slow ones form "the stability landscape," within which the lower-level systems move around, reacting to the changes in the "topography." The fast variables are crucial for the maintenance and increasing resilience of the system. They are often stimulated by disturbances, creating opportunities of renewal and novelty within the adaptive cycle of exploitation-conservation-renewal-reorganization (Holling 1995). Also, the maintenance of a basic capital or storage is important for keeping a high resilience. The resilience of a system changes as it goes through the phases of the adaptive cycle. The conservation phase (the climax), characterized by high niche-packing and high biomass, is very stable but brittle. It is a local and narrow stability domain and the resilience is low. The exploitation phase, on the other hand, is not very stable, but it has a high resilience.

Loss of resilience is caused by, for example, slow structural changes such as changes in the stability landscape, loss of novelty and variability, and accumulation of large storages. Our analysis indicates that the resilience of the Baltic Sea is to a large extent due to its large size and heterogeneous structure. The internal cycling in each one of the three systems can be described as an infinity loop or Holling four-box model, containing succession, breakdown, and regeneration. The individual loops are, however, strongly coupled (figure 4.10).

Key Variables

Slow variables are represented by, for example, the nutrient storages, especially phosphorus, which are successively accumulating in the deep bottom sediments (since the only other escape mechanisms are fish catches and fossilization in the sediments). Another variable, which has been changing for a long time but has been fairly stable during the last decades, is the nitrogen/phosphorus ratio, maintained at low levels through large-scale denitrification in the sediment bottoms. The pool of particulate organic matter is another example of an accumulating storage, which is important for the stability of the Baltic as well as of most ecosystems (Odum 1983). But it is also a sign of decreasing resilience.

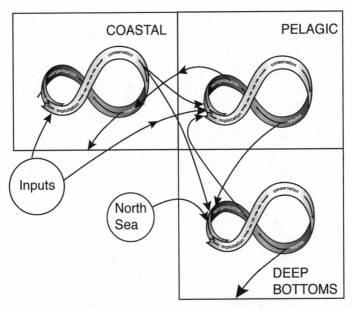

Figure 4.10. Systems turnover and systems couplings. Export-import between the coastal and pelagic system, sedimentation to deep bottoms from coastal and pelagic areas, and the nutrient feedback to the surface layer are connecting the cycles of respective ecosystem functions *sensu* Holling (1995).

Fast variables and key elements of the dynamics within the system are the phytoplankton and annual seaweeds, the microbes, and pelagic, particulate matter.

Disturbances are of many different kinds and magnitudes, triggering change in the ecosystem. The small-scale disturbances are incorporated into the system, and they contribute to species diversity by forcing communities back in the succession cycle. This means that at larger spatial scales a spectrum of different stages are maintained, where *r*- and *K*-strategists are living together at the same time. Several of the disturbances are meteorological ones, such as the influx of freshwater run-off, which, when increasing, triggers fish migration and stimulates the system to pulses of high primary production. Storms in early spring, which reshuffle the loose sediments, are important for washing off the cover of detritus settled on the primary producers under a sheltering cover of fast ice and may also influence the peak of the spring bloom. A similar crucial period occurs when the first autumn storms clean the cover of filamentous epiphytes from the seaweeds, which were produced during May–July, and the fragile algae, accumulated on shallow bottoms, are successively transported offshore, where they constitute an important input of organic matter to the deep soft-bottom system.

Ice scour is another physical disturbance that varies in strength between years. The scraping off of perennial algae from the shallow hard bottoms exposes them to colonization of spores of fast-growing annual algae and larvae of sessile animals such as blue mussels. "Strong winters give green years and mild winters brown years" is a proverb describing the switching dominance between the annual green algae, which are freshwater species and therefore favored by higher precipitation, and the marine brown algae. These are examples of opportunities for novelty during the reorganization phase at the start of the year.

Finally, the North Sea pulse, described above, often connected to westerly storms, is a large-scale disturbance, greatly affecting the deep bottom communities.

Biodiversity is a basic systems property, which in the Baltic case must be highlighted in connection to the resilience concept. The Baltic has a low species diversity of macroscopic animals, on the order of one-tenth the species diversity of a fully marine system. The organism assembly has had a short time to evolve. The Inland Ice left southern Scandinavia some fifteen thousand years ago; since then alternating periods of limnic, marine, and brackish water have created a fairly rich mixture of freshwater and marine organisms, which existed until the industrialization in the drainage basin accelerated and started to push the system from the center of its stability domain. Two views of the stability/diversity concept have been put forward. According to Jernelöv and Rosenberg (1976), the species diversity in the Baltic is certainly low, but it is maintained by hardy species, evolved in coastal areas with fluctuating conditions, and therefore robustly resistent to changes. Others have argued that the number of species in the different guilds, which maintain the basic ecosystem functions, is so low that an increased stress, killing off both individuals and species, decreases the buffering capacity and resilience of the system (Jansson 1980; Tedengren and Kautsky 1986).

There are few examples in the historical time of species that have been totally lost to the Baltic system, while some thirty immigrants have been recorded since the Viking Age (Elmgren and Hill 1997). Some of them have been shown to be crucial for maintaining basic processes in the system; others are still too new and scarce to have established any clear relations to the rest of the system. Some are regarded as real threats with potential to cause large-scale changes: the recently arrived zebra mussel and the American comb jelly *Mnemiopsis leidyi*. The latter has invaded the Black Sea, where it in huge numbers is vacuum-cleaning the waters of zooplankton and fish larvae (Mee 1992) and acting as a potential creative destructor of the pelagic community.

Reversible Changes: Fluctuations between Stability Domains

The different structures and time scales of the three subsystems of the Baltic make them examples of different types of resilience, although this is not easy to prove due to lack of historical data.

THE COASTAL SUBSYSTEM

This system has a varied topography with a mixture of soft and hard bottoms. In its pristine state it was characterized by submerged vegetation offering a rich number of niches to the organism assembly, but due to the change of turbidity it has been transferred to an alternative state dominated by annual, filamentous algae growing on all kinds of hard surfaces. It took half a century of increased nutrient loading for the nutrients and soft sediments to reach the level when the filamentous algae started their vigorous growth (in the 1960s) and the bladder-wrack retreated. From the sparse data available, one can presume that in the middle 1970s, when the filamentous brown algae suddenly exploded (Jansson and Kautsky 1977), the system was close to the bifurcation point between a perennial vegetation-dominated state and a state dominated by filamentous annuals. The increasing nutrient loads, mainly from an expanding use of nitrogen fertilizers in the 1960s–1970s, have driven the system deeper into this domain. There are, however, examples of local redevelopment of the perennial vegetation in areas where the nutrient loading from point sources has been reduced. The potential of the system to flip back to the previous state demonstrates the existence of alternative states, with no loss of hard stability so far. A similar switching has been observed in European lakes: flipping between a clear water state with submerged vegetation and a plankton-dominated state with low water transparency and no higher vegetation (Scheffer et al. 1993). There is a difference in the time scale, however, which might affect the classification as alternating or alternative system. The nutrient dynamics of Baltic coastal systems, like other marine coastal areas, are governed by the large offshore pool of nutrients and other critical elements with long turnover times. The recovery period is therefore much longer than that of a lake and needs for its quantification a wealth of long-term, historical data. The Baltic coastal system seems to be an example of an "excitable system" (Ludwig et al. chapter 2, 1997) where a recovery after a relaxed disturbance takes a long time.

Loss of a keystone species like *Fucus* would certainly mean a loss of hard stability (Ludwig et al. chapter 2). A future system without *Fucus* will be a detritus-based system with annual, filamentous seaweeds and higher plants on the soft bottoms, restricted to shallow areas because of the low water transparency. The biodiversity will be lower both in species, functions, and habitats.

The depth penetration of the *Fucus* can therefore be identified as an indicator of resilience, registering how far from the center of the stability domain (defined as the position before the 1940s) the system is working. Because of its intricate reproductive behavior, triggered at full moon during calm summer nights, and because its spores depend on clean surfaces for their settling, Serrao et al. (1996) have postulated that it will take a long time for the *Fucus*-dominated perennial state to return. Recent studies have shown, however, that there is a small and geographically patchy *Fucus* population in the Baltic reproducing in the autumn, when the cover of the annuals on the rocky bottoms has disappeared. This exemplifies the importance of genetic diversity for the resilience of the system. Will evolution favor a future dominance of autumn-spawning bladderwrack and thereby facilitate the return of the *Fucus* community?

OFFSHORE PELAGIC SYSTEM

The permanent halocline and the hydrodynamic border toward the coastal subsystem makes the offshore pelagic subsystem of the Baltic Proper look like a box with a leaking bottom and sides open to a large-scale, southbound throughflow. The Baltic Proper is mainly forced by light, nutrients, and temperature; it exports its solid wastes to the deep basins, its water of low salinity to the North Sea, and has no shortage of oxygen.

The present low nitrogen/phosphorus ratio and the increased nutrient levels have pushed the previously ordered pelagic system toward a more noisy state. Displacement in the pelagic foodweb originally caused by the increased nutrients involves both the lower trophic levels—shifts in phytoplankton and zooplankton—and the top consumer level—a decrease of the cod population, causing an increase of herring and sprat populations. Several examples of such changes have been given earlier: the large amount of sedimenting diatoms (overexploiting the water column of silica and starting to decrease); the decrease in surface salinity (favoring cladocerans but reducing the larger copepods important as fish food); and the increase in blooms of cyanobacteria ("blue-green algae") and other previously uncommon, toxic algae.

THE DEEP BASINS

The deep bottom flips between two stability domains: an oxygenated state with food webs incorporating higher organisms (including fish) and an anoxic state dominated by sulfur bacteria and hydrogen sulfide formation. The flip can be described as the shift between two attractors in a state space where oxygen forms the valley in the basin of attraction, which becomes shallower toward the bifurcation point, in the terminology of Hansell et al. (1997). Changes in the oxygen content would then drag the system up and down this valley. A repellor on the rim of the basin would be the isocline of 2 milliliters of oxygen per liter, which is the approximate limit for the presence of fish.

Irreversible Changes

There are few examples of extinction of species in the Baltic ecosystem; instead, immigration has increased the species diversity. What this means for the stability is not clear. Loss of whole key functional groups has not occurred but the potential effect of such incidents may be exemplified by the earlier described loss of silica, decline of diatoms, and decreasing diatoms/xanthine formation. Large-scale disappearance of *Fucus* would be close to irreversible and would in any case change the basic structure of the system.

The accumulation of phosphorus in the bottom sediments is not irreversible per se, as inorganic phosphorus is released to the bottom water in anoxic conditions. Holm (1978) reported on emptied phosphorus pools in the Bornholm Deep after long stagnation periods. But in oxygenated sediments the increasing and transformed loads of phosphorus are stored in apatite form. Is this to be compared to the immense amounts of excess carbon, fossilized during earlier geological periods, as it was not needed by the former systems? Will someone in the future try to exploit these submersed phosphorus mines when phosphorus has become a limiting mineral?

People and Issues of the Baltic

So far the description has been concentrated on the basic resilience of the Baltic marine system, although the effects of the drainage area and the human population have been included in the dynamics of the main state variables. Humans are an integrated part of the ecosystem, but compared with other animal species their incredible capacity to transform nature makes it necessary to distinguish a separate human system. Bearing in mind that a sustainable development is based on the resilience of the total system, one must ensure that the exploitation of various natural resources is kept within the tolerance limits of the ecosystems. This section characterizes the relationships between the human and the natural system, but a fuller treatment of the socioeconomic and political issues and their integration with the natural system requires transdisciplinary research, an area which requires far more attention than that given here.

Figure 4.11 shows some major couplings between humans and nature in the Baltic drainage basin, here divided into three interconnected sections. Appearing in figure 4.11 are, from left to right, (1) the main driving variables or large-scale drivers; (2) the sea and its drainage area consisting of riparian land and

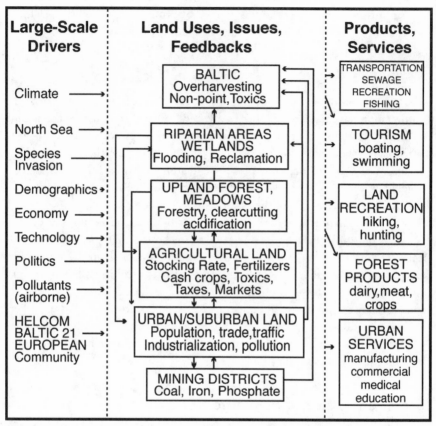

Figure 4.11. Diagram of the external driving forces that influence environmental issues and ecological services of the Baltic Sea in context of the larger system of Baltic Europe—the sea and its drainage area. The driving forces are the same for the drainage area and the enclosed sea. Land conversions push the sea around in its present stability domain and affect the water quality and the products of the Baltic Sea—especially fish and tourism.

wetlands, upland forests and meadows, cultivated land, urban areas, mining districts, and (3) the ecological and human-generated services and products essential for the survival of the total system. Feedback flows of energy, matter, people, and information connect landscapes, seascapes, and urban areas.

Large-Scale Drivers

Among the driving forces, three represent the natural system: climate, the North Sea, and species migrations and invasions. While demography, economy, and the use of fossil energies constitute the main forces of the human society, the byproducts of resource exploitation in the form of "pollution" are also added from outside through airborne flows. To control the exploitation and bring the total system closer toward sustainability, new institutions have been created. Three important ones are represented here, all linked to the global network of environmental institutions: the Helsinki Commission (HELCOM), Baltic Agenda 21, and, finally, the European Union, which plays an increasing role in regulating the economic transactions in the Baltic region.

Land-Use Patterns and Impact Issues

Fundamental for the resilience of the Baltic Sea is the land use in the drainage basin. Activities influencing the flows of nutrients and toxic substances are crucial in pushing the system around within—and even across—the borders of its stability domain. In this larger system, the Baltic Sea functions as a huge sink for waste products from resource use in the drainage area. The riparian and wetland areas act as buffer zones between land and water systems and are subject to changes through land reclamation, hydropower generation, and other forms of exploitation. A GIS (geographic information system) database of land cover and population density in the drainage basin, presented by Sweitzer et al. (1996), makes it possible to correlate the observed changes of the Baltic Sea with changes in land use.

Forested land is the dominant landscape type in the north, covering about half of the land area of Sweden and Finland. For centuries, forestry has been a basic sector of the economy. The race for higher profits has resulted in large-scale clearcutting and consequent increased run-off of nutrients. Pulpmill effluents have decimated marine bottom fauna and caused skeletal deformation in fish and seals.

Agriculture dominates the southern part of the drainage basin and covers 60–70 percent of Germany, Denmark, and Poland. During the Soviet era, Estonia, Latvia, and Lithuania were also large producers of agricultural products for consumption within the USSR. Since Soviet occupation ceased, the collective farming system has been replaced by small, private farms, and the use of artificial fertilizers and herbicides has decreased in the eastern countries because of a declining market and shrinking economies. The future of the Baltic Sea depends to a large extent on the form of economic development, especially in Poland. Efforts to reform its old-fashioned farming system into an energy-intensive western-type agriculture would have substantial, negative impacts on the sea due to the large share of the drainage basin and increased leaking from the sandy soils of Poland.

The expanding urban centers of industry, commerce, and administration, especially along the coasts of the Baltic Sea, act as large "consumers" of energy and material, and they export large amounts of wastes, trusting limitlessly "the self-purifying ability of nature." They have also created a fragmented landscape through their need for rapid communication and access to a vast network of transportation routes.

Ecological and Economic Services

The main ecological services of the Baltic Sea for society have already figured in the previous description of the marine ecosystem. The economic services (major ones are listed in figure 4.11) have long been regarded as outside the ecological system, following their own dynamics, inevitably testing the resilience of the ecosystem. Few studies have tried to penetrate the true couplings of the economic and ecological services, such as the freshwater flow for ecosystem services (Jansson et al. 1998) and the production of fish, and their dependence on the ultimate resource base, the solar-energy fixing primary production. One early, ambitious effort was the study of the island of Gotland in the Baltic Sea, where both the marine and terrestrial parts of this mesoscale ecosystem were mapped, analyzed, and synthesized on the basis of its energy and economic flows (Jansson and Zucchetto 1978; Zucchetto and Jansson 1985). Historical data show a total reliance on solar-based services, with expanding agriculture and forestry after the Bronze Age. Surplus production constituted the base for a flourishing trade when shipbuilding skills were acquired; thereafter the fish stocks, especially herring, could be tapped more effectively. The carrying capacity for humans was severely strained during the eighteenth and nineteenth centuries when the growing population increased the demand for food, housing, firewood, and other essential goods (Jansson 1985). The import of fossil energy rescued Gotland's forests from further over-exploitation, but the transformation of land continued with increasing agriculture, import of artificial fertilizers, reclamation of wetlands, and overuse and pollution of the precious groundwater. The latter was further affected by the quarrying of limestone to satisfy society's growing demand for cement. After food production and tourism the cement industry has the highest groundwater consumption. The overconsumption of water is probably the heaviest strain on the resilience of this system and is indeed a growing problem in certain parts of Baltic Europe. The shadow price of groundwater in the 1980s was nearly four times the price paid for water by industry on Gotland, which is one quantitative measure of the imposition of consumers on the life-support functions of the ecosystem. The historical description of Gotland, showing the switch from a solar energy–based human society, mainly depending on local resources, to a fossil fuel–based, highly technological society, mainly affected by fluctuations in the external market and transboundary flows of energy and materials, is probably valid for most Baltic coastal areas, although in various time scales. Trends for the whole of Baltic Europe are shown in figure 4.12.

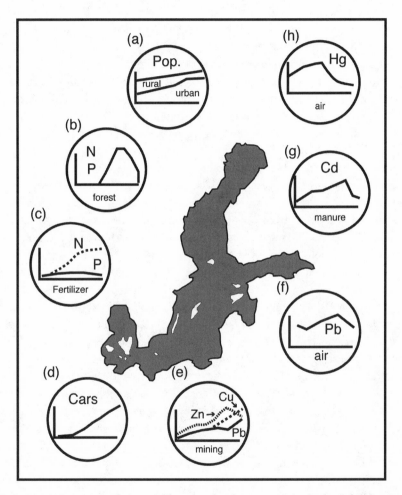

Figure 4.12. Recent trends in variables related to human activity in the Baltic Sea area between 1940 and 1990. (a) urbanization in Sweden; (b) fertilization in Sweden per 1,000 hectares; (c) trends for nitrogen and phosphorus in the use of artificial fertilizers in Sweden; (d) number of cars in Denmark, Finland, Sweden, Poland, Germany, and Great Britain (maximum of 70 million); (e) emissions of zinc, copper, and lead from mining in Sweden; (f) air emission of lead in Sweden; (g) emission of cadmium from manure; and (h) air emission of mercury.

The contribution of ecological services to the economy can be expressed as the life support value of the ecosystem. However, due to a rich supply of fossil fuels and minerals, the economy's dependence on ecosystem services has been masked. The successive displacement of natural capital with human-made capital is obvious in the fisheries. The increase in cod stocks in the 1970s and 1980s initiated big investments in larger fishing boats, equipment, and cod fillet factories (Folke et al.1991). Since then the catch per unit effort has decreased substantially due to the declining fish stock. The overcapitalized fishing industry suffers from big, structural, and economic problems and has become subject to strong regulations through an international quota system.

Another striking example of redirecting life support from natural to human systems is the decline of the Baltic grey seal population. At the beginning of the century some 100,000 grey seals were consuming about 255,000 tons of fish per year, which corresponds to about 65 percent of the total solar energy fixation area of the Baltic (Folke et al. 1991). Up to 85 percent of the total primary production of the Baltic Sea now supports the Baltic fisheries, when the necessary indirect energy flows from other parts of the system are included.

Around the 1950s humans and seals were about equally strong fish consumers. Seal populations were reduced from hunting and unintentional poisoning with PCBs, increasing the percentage take by humans. Hunting and pollution reductions led to an increase in the number of seals and an increase in competition for fish. Human fishers now want economic compensation and the right to hunt seals.

The methods of tapping terrestrial primary production through agriculture have also changed drastically since World War II due to low prices for oil and artificial fertilizers. Through increased mechanization and lavish use of artificial fertilizers the efficiency of crop production has increased. Farms have decreased in number but increased in size. Large units of cultivated land have favored the use of covered drains. This means that the important denitrification service of the traditional network of open ditches in the agricultural landscape has been transferred and forced upon the receiving areas of lakes and the Baltic coastal waters. Together with the effluents from the coastal sewage plants this means that the Baltic Sea now has an increasing role as decomposer of organic material originally produced outside the marine system and therefore without any natural feedbacks to the input environment.

The reclamation of wetlands has increased the leakage of nutrients to the marine system, decreased the groundwater storage, and decreased the buffering capacity of the landscape toward storm floods. In Poland, where probably 90 percent of the wetland areas have been reclaimed (Rydlöv et al. 1991), the potential risks of this development were drastically demonstrated during the flooding of the rivers Oder and Vistula in July 1997. Ten days of rain in Poland caused a freshwater outflow from the two rivers, corresponding to a week's normal outflow from the whole drainage area (Brandt et al. 1998). Twenty-seven cities and nearly two hundred villages were flooded. Due to prevailing hydrological conditions during this event, most of the nutrient-rich river water stayed in the coastal lagoons where a ten-fold increase in primary production occurred (Vedin et al. 1999); the larger Baltic system was only slightly affected.

Forests represent one of the major natural resources of the Baltic region. The forest industry has undergone rapid development due to high world demands for forestry products. The strategy has been to promote fast growth and high yields of timber, neglecting the important ecological services performed by the forests in watershed protection and maintenance of a high biodiversity. In Sweden, the production of pulp and paper has increased more than three and six times respectively since the 1950s. In the 1970s and 1980s alarming reports of increased acidification of lakes and forest soils in Scandinavia due to atmospheric deposition of sulfur and nitrogen oxides from coal mining and burning of fossil fuels in the heavy industrialized areas in Central Europe and Great Britain gave rise to a heated, international debate about the environmental costs of transboundary pollution. In fact, the increased downfall of nitrogen caused increased production in the northern coniferous forests while many forests in the southwest were severely damaged from acidification and nutrient losses. Increased knowledge about the relationships between the growth and management of forest resources, pollution control, and economic development has led to more environmentally friendly processing techniques and ecologically adapted harvesting practices. The forest industry is successfully adjusting to the long-term capacity of its living resource base, but how this will affect the species diversity of the forest flora and fauna is still an open question.

Urban services have developed from small-scale marketing of goods produced from local resources in a labor-intensive hinterland to the large-scale manufacturing of products for the international market. Many old towns in the Baltic are located at the border between the land and the sea, often at river mouths. The first period of exponential growth was the activity of the Hanseatic League during the thirteenth and fourteenth century, trading the Baltic products of fishing, agriculture, and forestry in the east and creating flourishing towns such as Lübeck, Rostock, Gdansk, Wisby, Tallinn, Riga, and Novgorod. Later, the expanding growth of the urban areas along the coasts was fueled by access to new energy sources (coal, oil, and electricity) and to new inventions like the steam engine and the internal combusion engine. Through the expansion of shipping, railways, and road networks, the towns and cities mushroomed over the Baltic Region with big cities such as St. Petersburg, Stockholm, and

Copenhagen situated on the coasts. These consumer systems must depend on large inflows of material and large outflows of degradation products. Folke et al. (1997) calculated the total ecological footprint (Rees 1992) or life-support area of the twenty-nine largest cities of the Baltic drainage basin. They found that the area for resource consumption and absorption of wastes amounts to 75–150 percent of the total drainage basin. This consumption by roughly 25 percent of the total population of the region also indicates that the indigenous natural resources of the Baltic area are overexploited and that the future development of the Baltic is governed by forces outside the total system. In reality, there is a strong dependence on imports of material from foreign areas and export of pollution.

The growth of the urban populations has promoted the increase of a new "industry"—tourism. The general improvement in living conditions for human societies built on energy-efficient, less labor-intensive production creates a demand for recreation, where travel and outdoor life are important components. The diversity of the coastal system, in habitats and scenery, species of flora and fauna, and cultural assets such as buildings and customs, makes it especially attractive. Along with swimming, sunbathing, and boating, sportfishing and hunting have become popular activities. Protecting the necessary life support for the favored wildlife targets, including high species diversity, thus becomes an important prerequisite for the tourist industry. The preservation of still-pristine areas of the Baltic has acquired economic value in addition to basic ecological functions, which may be easier for the layman to understand.

The Baltic archipelagoes have a high value for tourism and are at the same time examples of old, low-energy settlements to a large extent still based on solar energy. They constitute excellent subjects for principal studies of the resilience of a combined natural-cultural system. An important feature is the pulsing character of the system (Jansson and Hammer 1999) going back to the pattern of solar insolation and its derivatives of light and temperature, which have formed the annual rhythm of the archipelago society. In absence of fossil energy sources, the islanders had to adjust their small-scale fishing and farming to the natural, annual pulse. The resilience of the Baltic system was never really tested—there was a natural, sustainable relationship between man and nature. However, the harsh conditions in these coastal areas made people emigrate from the archipelago to the mainland. Island farms nearshore were abandoned and the absence of grazing cattle led to reforestation of land, while fishing diminished and was replaced by offshore fishing from urban ports. Since the 1950s, the archipelagoes have changed from livelihood to recreational areas dominated by summer houses and boat traffic (Andersson and Eklund 1998). The invasion of tourists leads to overexploitation of the ecosystems, especially the freshwater reservoirs and the stationary fish species.

The transformation of the archipelago is typical of many rural areas in the Baltic region. It has affected the rhythm, intensity, and form of natural resource use, with a consequent mismatch between resource availability and human demand. It is clear that to reestablish a sustainable relationship between humans and nature in the archipelago one must consider not only the couplings within its subsystems but also the larger system dynamics.

Governance and Resilience

The sustainable use of the Baltic Sea is a severe challenge that exhibits most of the problems of governing a multipurpose, common property resource. It involves a diverse set of stakeholders: societies of different countries with different customs, languages, histories, and political ideologies. Much of the research needed for dealing with these complex problems still remains to be done. Here we can only point at some initial steps of this governing process, mainly focusing on the basic resilience of the marine ecosystem—the Baltic Sea.

An early example of successful governance on the local scale is the fairly intricate annual rotation pattern of fishing lots in medieval communities, showing a clear understanding of the dynamics and carrying capacities of fish populations for human harvesting. Exploitation of fish resources emerged at the regional scale due to improving harvesting and shipping techniques and the export of salted herring in the fifteenth and sixteenth centuries (Odén 1980). The decline of the Hanseatic League is said to be at least partly due to the decline of the herring stock, probably an example of natural fluctuations. Fisheries of today have reached other significant dimensions, however. The ten-fold increase of catches between 1950 and 1970 led to fear of overfishing and to a need for international cooperation. An international agreement for the Baltic fishery was signed in 1973 by six of the littoral states, and later by all countries in the region.

In the 1970s, the marine pollution of the Baltic Sea gained worldwide attention. The detrimental impacts of toxic substances, nutrients from rapidly growing industries, and urban sewage systems came into focus and have since largely dominated the management issue (for a historical treatment see Jansson and Velner 1995). Concerted regional efforts were evoked by the Stockholm Conference in 1972. An international agreement to regulate waterborne and airborne pollution from land and ships was signed in 1974, and the Helsinki Commission (HELCOM) was established as the major governing institution. The political problems were severe and an extension of the agreement to include the entire catchment area had to wait until 1992–1993 after the breakdown of the Soviet Union.

At that time, the governments of the Baltic countries, shocked by the Chernobyl accident in 1986, heavy blooms of blue-green algae, and crashing cod stocks, had recognized the large-scale character of the stresses on the systems resilience. In 1988, the environmental ministers announced a 50 percent reduction of the discharges of nitrogen and phosphorus. A large number of "hotspots" of industrial, municipal, and agricultural polluters became subject to a joint comprehensive action program. This international cleaning effort was planned to be carried out in collaboration with four international banks—the first agreement of its kind, signed by the prime ministers of the nine Baltic countries. In 1996, the ministers also decided to develop an Agenda 21 for the Baltic Region (Kristoferson and Stålvant 1996), wherein the individual countries were given the task of preparing scenarios of sustainable resource uses for the main sectors of the community: agriculture, forestry, industry, transportation, energy, and tourism.

Concluding Remarks

With its past activities, society has driven the Baltic Sea ecosystem toward an irreversible stable state characterized by decreasing blooms of diatoms, increasing blooms of nitrogen fixing and sometimes toxic cyanobacteria, filamentous algae outcompeting the perennial keystone *Fucus* alga, and a fish community dominated by scrapfish. Increasing oxygen consumption by the seafloor increases the periods of anoxic deep bottoms and the unfavorable conditions for the reproduction of the commercially important cod. The system is losing its buffering capacity, sliding toward the borders of the present stability domain. Susceptible to disturbances, the system has low capacity of incorporating the effects of surprises. Rehabilitation of the previous natural state with a flourishing seal population is combated, and Baltic Europe drives itself deeper into a dependence of a fossil fuel-based, fragile human-nature system.

The development during the 1900s can be classified into three periods. Stage 1, the period before 1940, was an oligotrophic stage with clear water, low primary production, luxuriant growth of macroalga *Fucus*, and large stocks of commercial fish, including the wild salmon. Stage 2, 1940–1980, exhibited increasing nutrients, decreasing water transparency, decreasing *Fucus*, and increasing filamentous algae, decreasing wild salmon, and increasing reared stocks. Oligotrophy changed to eutrophy. Stage 3, 1980–1999, shows nutrients, primary production, and macroalgae leveling off, cod decreasing, and wild salmon close to extinction and replaced by reared salmon.

In terms of resilience, a simple classification of the total Baltic Sea falls back on the individual subsystems and their role in the higher system. The coastal subsystem is close to a bifurcation of a high-diversity, clear-water system dominated by perennial seaweeds and a low-diversity, turbid system with annual, filamentous alga. It looks like an "excitable system," which needs a long period of recovery after relaxed disturbance. Loss of hard stability has not yet occurred. The offshore pelagic system is changing toward a more "turbulent" state, where the previous stable pattern of food-web structure and dynamics adopt different size and shape of its meshes. Algal blooms change in size and quality. The deep soft bottoms offer a clearer example of flips between an oxic state of a detritus-based organism assembly with phosphorus accumulation and denitrification, and an anoxic state dominated by sulfur bacteria and phosphorus export. Couplings to the other subsystems is obvious in the Gulf of Finland, where transport of deep phosphorus from the Baltic Proper cause increased blue-green blooms and growth of annual seaweeds. Compared to previous stages, the present total Baltic shows a faster reaction to changes in solar insolation by increasing production because of the high nutrient loads already present in the system.

The total systems dynamics in the temporal dimension have been described as a battle on two fields. One is fought between the saltwater from the North Sea and the freshwater from land run-off. Directed both by the elevation of land after the last glaciation and by meteorological forces—the freshwater side also by human activities through regulating river flows through hydroelectric dams—the fight is moving back and forth. During the 1970s, the saltwater gained victories; during the 1980s–1990s, the freshwater dominated and meager plankton food was offered the pelagic fish. The other battle is fought on the deep bottoms between oxygen and hydrogen sulfide. Also, here the weather forcings are basic but human activities join the anoxic side through eutrophication by forestry, agriculture, and urban activities. The two wars are coupled and the combined result can be described as the oligotrophic state losing to the eutrophic side.

Originally conducting small-scale, locally directed resource management, the Baltic institutions have during the last thirty years developed a regional, large-scale governance of a stressed sea of limited resilience. The bold, ministerial declarations will have no practical effects, however, if they are not effectively operating at the local scale and include the behavior of the individual. Non-governmental organizations, of which some thirty have formed the Coalition Clean Baltic (CCB), play an important role, encouraging the public to a less-affluent lifestyle and putting pressure on the authorities to implement the adopted international agreements. Concerted actions are important tools for keeping our life-support systems at a functioning level.

Acknowledgments

We thank Lance Gunderson, Atlanta, Georgia, United States, for skillfully guiding us in the mist of details, and Ted Munn, Toronto, Ontario, Canada, for stimulating discussions and help with literature.

Literature Cited

Andersson, K., and E. Eklund. 1998. Outdoor recreation, nature protection and political opportunity structures. The case of the Finnish Archipelago Sea National Park. Paper presented at the workshop "Outdoor recreation—practice and ideology". Umeå, Sept. 2–6, 1998.

Aneer, G. 1987. High mortality of Baltic herring (*Clupea harengus*) eggs caused by algal exudates? *Marine Biology* 94:163–169.

Bengtsson, B.-E. 1998. The present status of the M74 Syndrome—Report from the Swedish FiReproject. Proceedings of the 1st Baltic Marine Science Conference, Rönne, Bornholm, Denmark. ICES.

Bergström, S., and B. Carlsson, 1994. River runoff to the Baltic Sea: 1950–1990. *Ambio* 23:280–287.

Bernes, C. 1988. *Monitor 1988.* Sweden's marine environment ecosystems under pressure. Stockholm: National Swedish Environmental Protection Board.

Brandt, M., L. Edler, and L. Andersson. 1998. Översvämningarna längs Oder och Wisla sommaren 1997 samt effekterna i Östersjön ("Flooding along Oder and Vistula in the summer of 1997 and its impact on the Baltic Sea") SMHI, Oceanography 68, Norrköping, Sweden (in Swedish).

BSEP (Baltic Sea Environment Proceedings). 1996. Baltic Sea Environment Proceedings No. 64 B. Third Periodic Assessment of the state of the marine environment of the Baltic Sea, 1989–93, Background Document. Helsinki Commission.

Cederwall, H., and R. Elmgren. 1980. Biomass increase of macrobenthic fauna demonstrates eutrophication of the Baltic Sea. *Ophelia* Suppl.1:287–304.

Elmgren, R. 1978. Structure and dynamics of Baltic benthos communities, with particular reference to the relationship between macro- and meiofauna. *Keiler Meeresforschungen* Sonderheft 4:1–22.

———. 1989. Man's impact on the ecosystem of the Baltic Sea: Energy flows today and at the turn of the century. *Ambio* 18:326–332.

Elmgren, R., and C. Hill. 1997. Ecosystem function at low biodiversity—the Baltic example. Pp. 319–336 in : *Marine biodiversity, patterns and processes*, edited by R. F. G. Ormond, J. D. Gage, and M. V. Angel. Cambridge: Cambridge University Press.

Flinkman, J., E. Aro, I. Vuorinen, and M. Viitasalo. 1998. Changes in northern Baltic zooplankton and herring nutrition from 1980s to 1990s: Top-down and bottom-up processes at work. *Marine Ecology Progress Series* 165:127–136.

Folke, C., M. Hammer, and A.-M. Jansson. 1991. Life-support value of ecosystems: A case study of the Baltic Sea region. *Ecological Economics* 3:123–137.

Folke, C., Å. Jansson, J. Larsson, and R. Costanza. 1997. Ecosystem appropriation by cities. *Ambio* 26:167–172.

Hansell, R. I. C., I. T. Craine, and R. E. Byers. 1997. Predicting change in non-linear systems. *Environmental Monitoring and Assessment* 46:175–190.

Hansson, S., and L. G. Rudstam. 1990. Eutrophication and Baltic fish communities. *Ambio* 19:123–125.

Hinrichsen, U., and F. Wulff. 1998. Biogeochemical and physical controls of nitrogen fluxes in a highly dynamic marine system—model and network flow analysis of the Baltic Sea. *Ecological Modeling* 109:165–191.

Holling, C. S. 1976. Resilience and stability of ecosystems. Pp. 73–92 in *Evolution and consciousness: Human systems in transition*, edited by E. Jantsch and C. H. Waddington. Reading, Mass.:Addison-Wesley.

———. 1995. What barriers? What bridges? Pp. 3–34 in *Barriers and bridges to the renewal of ecosystems and institutions*, edited by L. H. Gunderson, C. S. Holling, and S. S. Light. New York: Columbia University Press.

Holling, C. S., D. W. Schindler, B. Walker, and J. Roughgarden. 1994. Biodiversity in the functioning of ecosystems: An ecological primer and synthesis. In *Biodiversity: Ecological and economic foundations*, edited by C. Perrings, C. S. Holling, B.-O. Jansson, and K.-G. Mäler. Cambridge: Cambridge University Press.

Holm, N. 1978. Phosphorus exchange through the sediment-water interface: Mechanism studies of dynamic processes in the Baltic Sea. Contrib. Department of Geology, University of Stockholm. *Microbial Geochemistry* 3:1.

Hook, O., and A. Johnels. 1972. The breeding and the distribution of the grey seal *Halichoerus grypus* Fab, in the Baltic Sea, with observations on other seals of the area. *Proceedings of the Royal Society of London* 182:37–58.

ICES. (International Council for the Exploration of the Sea). 1995. Report of the ICES Advisory Committee on Fishery Management, 1994. ICES Cooperative Research Report 210. Copenhagen, Denmark.

Jansson, Å., C. Folke, and S. Langaas. 1998. Quantifying the nitrogen retention capacity of natural wetlands in the large-scale basin of the Baltic Sea. *Landscape Ecology* 13:249–262.

Jansson, A.-M. 1985. Natural productivity and regional carrying capacity for human activities on the island of Gotland, Sweden. In *Economics of ecosystem management*, edited by D. O. Hall, N. Myers, and N. S. Margaris. Dordrecht, Netherlands: Dr. W. Junk Publishers.

Jansson, A.-M., and M. Hammer. 1999. Patches and pulses as fundamental characteristics for matching ecological and cultural diversity: The Baltic Sea archipelago. *Biodiversity and Conservation* 8:71–84.

Jansson, A.-M., and N. Kautsky. 1977. Quantitative survey of hard-bottom communities in a Baltic archipelago. Pp. 359–366 in *Biology of benthic organisms*, edited by B. F. Keegan, P. O. Ceidigh, and P. J. S. Boaden. Oxford: Pergamon Press.

Jansson, A.-M., and J. Zucchetto. 1978. Energy, economy and ecologic relationships for Gotland, Sweden: A regional systems study. *Bulletins from the Ecological Research Community/NFR*, Stockholm 28.

Jansson, B.-O. 1978. The Baltic: A systems analysis of a semi-enclosed sea. Pp. 131–184 in *Advances in oceanography*, edited by H. Charnock and G. Deacon. New York: Plenum Press.

———. 1980. Natural systems of the Baltic Sea. *Ambio* 9:128–136.

Jansson, B.-O., and C. Källander. 1968. On the diurnal activity of some litoral per-acaride crustaceans in the Baltic Sea. *Journal of Experimental Marine Biology and Ecology* 2:24–36.

Jansson, B.-O., and H. Velner. 1995. The Baltic: The sea of surprises. Pp. 292–372 in *Barriers and bridges to the renewal of ecosystems and institutions*, edited by L. H. Gunderson, C. S. Holling, and S. S. Light. New York: Columbia University Press

Jansson, B.-O., W. Wilmot, and F. Wulff. 1984. Coupling the subsystems: The Baltic Sea as a case study. Pp. 5494–5595 in *Flows of energy and materials in marine ecosystems*, edited by M. J. R. Fasham. New York: Plenum Press.

Jensen, S., A. G. Johnels, M. Olsson, and G. Otterlind. 1969. DDT and PCB in marine animals from Swedish waters. *Nature* 224:247–250.

Jernelöv, A., and R. Rosenberg. 1976. Stress tolerance of ecosystems. *Environmental Conservation* 3:43–46.

Johnels A. G., T. Westermark, W. Berg, G. Persson, and B. Sjöstrand. 1967. Pike (*Esox lucius*) and some other aquatic organisms in Sweden as indicator organisms of mercury contamination in the environment. *Oikos* 18:323–333.

Jonsson, P. 1992. Large-scale changes of contaminants in Baltic Sea sediments during the twentieth century. Ph.D. diss. Uppsala University, Sweden.

Kaasik, T. O. 1989. The geography of the Baltic region. Pp. 15–22 in *Comprehensive security for the Baltic: An environmental approach*, edited by A. H. Westing. London: Sage Publications.

Karhu, M., B. Håkansson, and O. Rud, 1995. Distribution of the sea-surface temperature fronts in the Baltic Sea as derived from satellite imagery. *Continental Shelf Research* 15:663–679.

Karjalainen, M. 1999. The effect of nutrient loading on the development of the state of the Baltic Sea—an overview. *Walter and Andrée de Nottbeck Foundation Scientific Reports*. Report No. 17. Helsinki, Finland.

Kautsky, N., H. Kautsky, U. Kautsky, and M. Waern. 1986. Decreased depth penetration of *Fucus vesiculosus* (L.) since the 1940s indicates eutrophication of the Baltic Sea. *Marine Ecology Progress Series* 28:1–8.

Kristoferson, L., and C.-E. Stålvant. 1996. *Baltic 21: Creating an Agenda 21 for the Baltic Sea region, Main Report*. Stockholm: Stockholm Environment Institute.

Larsson, U., R. Elmgren, and F. Wulff. 1985. Eutrophication and the Baltic Sea: Causes and consequences. *Ambio* 14:9–14.

Leppäkoski, E. 1975. Macrobenthic fauna as indicator of oceanization on the southern Baltic. *Merentutkimuslait. Julk./Havsforskningsinst. Skr. No.* 239:280–288.

Ludwig, D., B. Walker, and C. S. Holling. 1997. Sustainability, stability, and resilience. *Conservation Ecology* [online] 1:1–8 http://www.consecol.org/vol1/iss1/art7.

Matthäus, W. 1995. Natural variability and human impacts reflected in long-term changes in the Baltic deep water conditions—a brief review. *Deutsche Hydrographische Zeitschrift* 47:47–65.

Matthäus, W., and H. Schinke. 1994. Mean atmospheric circulation patterns associated with major Baltic inflows. *Deutsche Hydrographische Zeitschrift* 46:321–339.

Mee, L. D. 1992. The Black Sea in crisis: A need for international concerted action. *Ambio* 21:278–286.

Odén, B. 1980. Human systems in the Baltic area. *Ambio* 9:116–127.

Odum, H. T. 1971. *Environment, power and society*. New York: Wiley InterScience.

———. 1983. *Systems ecology: An introduction*. New York: John Wiley and Sons.

Olsson, M., and L. Reutergård. 1986. DDT and PCB pollution trends in the Swedish aquatic environment. *Ambio* 15:103–109.

Rees, W. 1992. Ecological footprints and appropriated carrying capacity: What urban economics leaves out. *Environment and Urbanization* 4:121–130.

Rydlöv, M., H. Hasslöf, K. Sundblad, K. Robertson, and H. B. Wittgren. 1991. Wetlands—vital ecosystems for nature and society in the Baltic Sea region. World Wide Fund for Nature, WWF, Report to the HELCOM ad hoc Level Task Force.

Sandén, P., and B. Håkansson. 1996. Long-term trends in Secchi depth in the Baltic Sea. *Limnology and Oceanography* 41:346–351.

Scheffer, M., S. H. Hosper, M.-L. Meijer, B. Moss, and E. Jeppesen. 1993. Alternative equilibria in shallow lakes. *Trends in Ecology and Evolution* 8:275–279.

Schinke, H., and W. Matthäus. 1998. On the causes of major Baltic inflows—an analysis of long time series. *Continental Shelf Research* 18:67–97.

Segerstråle, S. G. 1957. The Baltic Sea. Pp. 751–800 in *Treatise on marine ecology and palaeoecology*. Vol. 1, *Geological Society of America Memoirs 67*, edited by J. W. Hedgpeth. New York: Geological Society of America.

Serrao, E. A., G. Pearson, L. Kautsky, and S. Brawley. 1996. Successful external fertilization in turbulent environments. *Proceedings of the National Academy of Science* 93:5286–5290.

Shaffer, G. 1975. Baltic coastal dynamics project: The fall downwelling regime off Askö. A report contrib. from the Askö Laboratory, Stockholm 26.

Shaffer, G., and U. Rönner. 1984. Denitrification of the Baltic Proper deep water. *Deep-Sea Research* 31:197–220.

Snoeijs, P. 1999. Diatoms and environmental change in brackish waters. In *The diatoms: Applications for the environmental and earth sciences*, edited by J. F. Stoermen and J. P. Smol. Cambridge: Cambridge University Press.

Sweitzer, J., S. Langaas, and C. Folke. 1996. Land use and population density in the Baltic Sea drainage basin. *Ambio* 25:191–198.

Tedengren, M., and N. Kautsky. 1986. Comparative study of the physiology and its probable effect on size in blue mussels (*Mytilus edulis* L.) from the North Sea and the northern Baltic Proper. *Ophelia* 25:147–155.

Thurow, F. 1989. Fishery resources of the Baltic region. Pp. 54–61 in *Comprehensive security for the Baltic: An environmental approach*, edited by A. H. Westing. London: Sage Publications.

Vedin, H., A. Eklund, and H. Alexandersson. 1999. The rainstorm and flash flood at Mount Fulufjallet in August 1997: The meteorological and hydrological situation. *Geografiska Annaler Series A, Physical Geography* 81A:361–368.

Voipio, A., ed. 1981. *The Baltic Sea*. Elsevier Oceanography Series, no. 30, Amsterdam: Elsevier Science.

Vuorinen, I., J. Hänninen, M. Viitasalo, U. Helminen, and H. Kuosa. 1998. Proportion of copepod biomass declines together with decreasing salinities in the Baltic Sea. *ICES Journal of Marine Science*. 55:767–774.

Wachenfeldt, T., S. Waldemarsson, and P. Kangas. 1986. Changes in the littoral communities along the Baltic Sea coasts. Baltic Sea monitoring symposium 1986. *Baltic Sea Environment Proceedings* 19:394–403.

Zucchetto, J., and A.-M. Jansson. 1985. *Resources and society: A systems study of Gotland, Sweden*. Berlin: Springer-Verlag.

5

Resilience of Coral Reefs

Tim R. McClanahan, Nicholas V. C. Polunin,
and Terry J. Done

Coral reef ecosystems are restricted to tropical latitudes (within 30° of the equator) and therefore experience relatively constant environmental conditions on the basis of the seasonal cycles. This relative stability may produce organisms and ecosystems that are poorly adapted to environmental fluctuations. It is, therefore, arguable that coral reefs may be among the least resilient ecosystems on the planet and may be good indicators of environmental changes such as global warming (Glynn 1993; Brown 1997). There is good evidence, however, that tropical environments and coral reefs experience significant intra- and interannual variation in environmental conditions, such as temperature and light (McClanahan 1988; Sheppard 2000), albeit less than temperate latitudes. Reefs are also one of the most persistent ecosystems over the Earth's geological history (Veron 1995). The species that have created reefs have, however, waxed and waned over time and the species and even taxonomic orders dominant in the present Cenozoic era are significantly different from those of the Mesozoic and Paleozoic eras (Veron 1995). Consequently, there has been considerable debate among reef ecologists and geologists over the issue of reef resilience and stability and whether the geologic past is relevant to the scale and types of contemporary disturbances (Brown 1997).

Coral are exposed to a large number of natural and human-influenced disturbances that affect their species and ecological adaptations and organization (Connell 1978; Rogers 1993; Hughes 1993; Brown 1997). Many of these ecological disturbances may, however, be averaged out, and reefs may appear to be more stable over geologic (Pandolfi 1996) than ecological time (Connell et al. 1997). Over geologic time (more than hundreds of years) the species composition of geologic deposits may remain fairly stable even in the presence of disturbances over ecological time (days to tens of years). In some cases species composition may change while total species richness remains similar, because newly evolved species can compensate for species extinctions (Jackson et al. 1993; Budd et al. 1996), but this may vary between ocean basins (Jablonski 1998). Gains and losses of species appear to occur on the scale of millions rather than of hundreds of thousands of years, and most extant reef species have been present over the past few million years or a number of glacial cycles (Veron 1995; Jackson et al. 1996).

Despite the potential long-term stability of reefs in the face of many disturbances, there is concern that recent human-induced environmental changes may be exceeding the limits of tolerance of reef organisms to factors such as water temperature, ultraviolet radiation, and predation by humans (Glynn 1993). For example, the warmest temperatures of the last 100,000 years were only about 1°C above today's. To find temperatures 3°C warmer, as predicted by global climate change models, one must look back several million years to the Pliocene, when the Earth's and coral reef biota were different from today's (Veron 1995; Livingstone 1996). These changes in background environmental factors will be associated with the rapidly rising use of reefs for fisheries and other types of resource extraction (coral building blocks, medicine, and so forth). Dual stresses could have variable consequences, with some counteracting each other and others being synergistic. Some synergistic relationships may lead to species losses and novel ecosystems with lost ecological services in cases where single stresses might have been tolerable.

Below we will introduce some of the basic aspects of reef ecology and discuss the conditions that appear to lead to characteristics of both resilience and non-resilience of ecological structures, functions, and their management at different spatial scales. Our analysis relies heavily on a number of recent case studies where reefs have or have not been resilient to environmental and human disturbances.

Reef Functions

One of the simplest ways to view the ecological function of coral reefs is through two of the main outputs of reefs—organic and inorganic (calcium carbonate) carbon production (figure 5.1). Carbon, largely available as bicarbonate ions dissolved in seawater, is fixed by reef organisms for the production, maintenance, and reproduction of the organisms themselves as well as their skeletal structures. The production of skeletal structure or the calcium carbonate production ranges from 1 to 10 kilograms per square meter per year (Kinsey 1985,

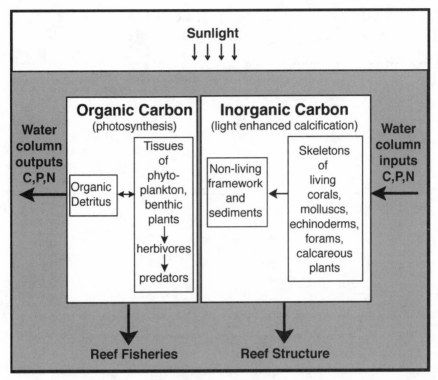

Figure 5.1. Diagrammatic representation of the pathways of carbon fixed in coral reef ecosystems. Diagram aggregates carbon accumulation by organic and inorganic (calcium carbonate) pathways.

1991). Skeletal structure is also broken down into sediments that form sandy bottoms and beaches and are an important fraction of the inorganic carbon pathway (figure 5.1). Gross organic production, as well, is also high at 1 to 9 grams of carbon per square meter per day, which is equivalent to 3.5 to 32.2 kilograms wet per square meter per year. Such production supports the complex and diverse organisms and food webs found on most coral reefs.

Primary production is largely dependent on a variety of algae that principally include four main functional groups. These are symbiotic algae living in hard corals (zooxanthellae), fast-growing filamentous turf-forming algae (turfs), larger fleshy red, brown, and green algae (fleshy algae), and algae that deposit calcium carbonate skeletons (red coralline and green calcareous algae) (Steneck and Dethier 1994). Each of these groups exhibits a slightly different balance between organic and inorganic production, or tissue growth versus skeletal or defensive structures. Corals and their zooxanthellae and coralline algae generally

Table 5.1. (a) Energy inputs into a typical coral reef system, and (b) standards of metabolic performance for three main types of benthic substratum

(a)

Energy Input	Actual Energy, joules/m^2/y
Solar energy	7.1E+9
Waves, absorbed	9.9E+8
Currents, kinetic	6.6E+8
Tides	1.1E+8
Rain, physical	3.1E+4

Source: Data from McClanahan 1990 based on East African data.

(b)

Substratum*	Photosynthesis gC/m^2/day	Production/Respiration	Calcification kg CaCO^3/m^2/y
Continuous coral	20	1	10
Algal pavement	5	> 1	4
Sand and rubble	1	< 1	0.5

Source: Data from Kinsey 1991.

* These three categories are the dominant substrata in the "framework," "pavement," and "sand" zones, respectively, of coral reefs. Varying proportions of one or both of the other two categories may be present.

have relatively low organic production but higher inorganic carbon production, while turfs and fleshy algae have the opposite pattern (table 5.1). As we will show later, reefs can be dominated by different functional groups that can greatly affect the ratios of organic to inorganic production.

Undisturbed coral reefs seldom seem severely nutrient limited. Field studies indicate that, as water passes over coral reefs, there is no net uptake of phosphorus and an export of nitrogen (Pilson and Betzer 1973; Wilkinson et al. 1984). This indicates that the physicochemical factors that influence production of both organic and inorganic carbon are primarily dependent on sunlight and water motion from waves, tides, and currents (table 5.1). Sunlight is the largest single energy input, but the total energy in water motion, of tides, currents, and waves combined is also large and is important in transporting resources and waste products of the reef. Perhaps equally important for maintaining high production are disturbances to the primary producers, which maintain an early stage of ecological succession and, therefore, high photosynthesis and growth rates of turf-forming and fleshy algae (Carpenter 1988; Choat 1991). Consequently, some investigators have argued that the abundance of grazers is the main limitation to primary production on coral reefs (Larkum 1983). Consumers, in general, frequently disturb their prey and they, therefore, may often be responsible for the high production of reefs by maintaining their prey in high-growth phases and by supplying concentrated nutrients to their prey (Polunin 1988; Meyer and Schultz 1985).

Ecological Services

Disturbances are important in maintaining high biological production and shaping general reef ecology. It should be recognized, however, that many ecological processes have a unimodal (hump-shaped) response to disturbance or production factors, and either too little or too much of the factor can reduce production or structure from a maximum level. It is the net production of resources that is available for humans, not the gross. Net productivity of both organic and inorganic carbon are considerably lower than gross measures because a significant portion of the production is lost from the reefs or, more importantly, consumed by the coral-reef organisms themselves, largely in respiration, self-maintenance, or reproduction (Polovina 1984; Birkeland 1997). In the case of inorganic calcium carbonate production, there are whole suites of organisms that erode these skeletal structures into sand (Glynn 1997). Sand can be exported from the reef to form sand-based ecosystems such as seagrass meadows and beaches. Consequently, as little as 1 percent of the gross organic production of reefs is available for humans, largely in the form of various seafood (Birkeland 1997). Few estimates of the amount of inorganic carbon available for human resources have been made, as humans do not directly use most inorganic carbon. Instead, inorganic carbon constitutes an indirect ecological service such as shoreline protection and the formation of beaches. Coral blocks from dead or living corals are important for building in many tropical coastal settlements, but little effort has been made to determine the sustainable level of extraction of calcium carbonate (Risk and Sluka 2000).

Alternate Structures and Interacting Processes

Coral reef ecologists have reported a variety of community or ecosystem states that are often described by the dominant organisms or functional groups (figure 5.2). These states are often attributed to human influences, where the two most common environmental concerns of reef ecologists have been the influences of

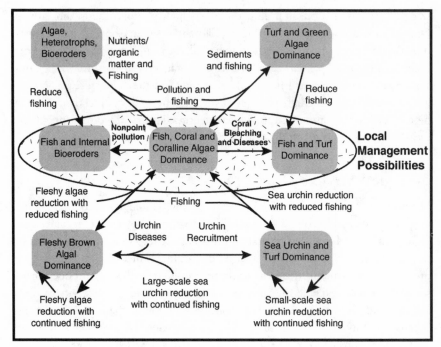

Figure 5.2. A simple conceptual model (meta-model), depicting the different ecological states (gray polygons) of coral reefs and the factors that may cause or maintain these states.

harvesting and pollution. A conceptual model based on major functional groups (figure 5.2) predicts a variety of ecological states, depending on combinations of pollution, fishing, sediments, diseases, and other natural disturbances such as cyclones and human management. Because reefs are often exposed to a combination of human and natural influences, individual reefs may be an amalgamation of many of these states.

Many reef ecologists believe that increases in nutrient concentrations in waters will switch dominance from corals to various forms of fleshy and filamentous algae and bioeroding sponges (Smith et al. 1981; Cuet et al. 1988; Bell 1992; Lapointe et al. 1997; Rose and Risk 1985; Risk et al. 1995). Nutrification, or the addition of growth-promoting nutrients such as nitrogen and phosphorus, has the potential to increase the growth rates of some fast-growing algae and heterotrophic invertebrates relative to corals (Littler et al. 1991; Lapointe et al. 1997). Many studies have shown, however, either only small or no responses to nutrient additions (Kinsey and Domm 1974; Kinsey and Davies 1979;

Hatcher and Larkum 1983; Larkum and Steven 1994; McClanahan et al. 2002). Some algae, such as browns, actually decrease in the presence of nutrification while others, such as greens and blue-greens, increase (Borowitzka 1972; McClanahan et al. 2002). In the Caribbean, some sites considered to be oligotrophic have high cover of macroalgae (e.g., Williams and Polunin 2001). All other variables being equal, an increase in nutrients and algal growth might be expected to cause a reef dominated by late-successional fleshy algae, sponges, and soft corals. In practice, however, there are a number of disturbances, particularly to algae, which prevent this from occurring. Most important, herbivory (Hatcher and Larkum 1983; McCook 1999), but also physical disturbances like water movement and the saltation of sediments along the seafloor, can frequently reduce algae and compensate for increased algal growth (McClanahan 1997a). Fishing, sometimes associated with a reduction in herbivory, and nutrification may interact to produce algal growth responses (figure 5.2).

The combination of fishing and pollution does not always shift reefs from dominance by corals and early successional algae to late-successional algae. For example, fishing can promote the abundance of herbivorous sea urchins (Hay 1984; McClanahan and Shafir 1990) that graze intensely and reduce the abundance of algae to levels even lower than when fish dominate grazing (McClanahan 1995a, 1997a). In a second example, when nutrient additions are associated with sediments from a river or a dredging operation, the physical disturbance per se can retard the successional development of algae (McClanahan and Obura 1997; McClanahan 1997a). Additionally, in areas or seasons with high wave or current energy, algae may be constantly disturbed and unable to maintain high levels of abundance (McClanahan et al. 1996). The abundance of macroalgal may also be increased by spatial escapes from grazing with the loss of coral cover through disease and other disturbances (Williams et al. 2001). Food webs are one way in which the functional linkages among ecosystem components are perceived, but because predation is only one process affecting abundance of organisms, their ability to predict indirect effects in ecosystems is very limited (Polunin and Pinnegar 2002). Consequently, there are a number of contingent outcomes to pollution and fishing disturbances depending on how these factors interact among themselves (McCook 1999) and with diseases and other physicochemical conditions.

In order to achieve the conservation objective of increasing the abundance and diversity of corals and fishes, managers have the option of trying to reduce fishing and pollution as well as reducing pest species or groups such as sea urchins and unpalatable fleshy algae. In many cases, it is important to attempt a combination of these management options as the outcome will often depend on an interaction of two or more factors (McClanahan et al. 1996, 2001).

Disturbances, Time Scales, and Reef Ecology

The kinds, scales, and intensities of disturbance vary both regionally and locally (Scoffin 1993), and may be modified by human population growth, resource exploitation, and industrialization. Episodic events (those lasting a few hours to months) that can injure, kill, and collapse most corals over scales of hectares include hurricanes (Woodley et al. 1981; Scoffin 1993; Rogers 1993; Massel and Done 1993), freshwater (Hedley 1925), predators (Endean 1976; Moran 1986), stress-related bleaching (Glynn 1993; Brown 1997), and sedimentation (Cortes and Risk 1985). Other disturbances, such as the many diseases reported in Caribbean corals in recent years, may take years to decades to decimate local population of corals (Antonius 1985; Aronson and Precht 1997). While we suggest elsewhere in this chapter that the quantity (frequency, intensity, and distribution) of each of these types of episodic and long-term disturbances has been modified greatly by humans, most disturbances, we believe, are qualitatively similar to disturbances reefs have experienced since their origin.

Natural Disturbances: Cyclones

The statistical likelihood of a particular coral reef being exposed to strong wave forces varies greatly. It is negligible in the doldrums, which extend to 10° north and south of the equator. Reefs between these latitudes are subject, instead, to weaker monsoonal storms. When and where extreme events do occur, the reefs involved are, at the very least, substantially set back to an earlier successional stage dominated by coral rubble and algal turf. At most, cyclones destroy the entire coral architecture and redistribute rubble and biogenic sediments (Woodley et al. 1981; Scoffin 1993; Dollar and Tribble 1993). Within cyclone latitudes, the expected "cyclone-free" longevity of a massive coral (Massel and Done 1993), which may also be thought of as a surrogate for the longevity of uninterrupted succession, first increases, then decreases with increasing distance from the equator. In some reef areas, the length of "cyclone-free" successional runs are predicted to shorten as cyclone frequencies increase under current global climate change scenarios (Pittock 1999; Done 1999).

Crown-of-Thorns Starfish

Populations of coral-eating crown-of-thorns starfish *Acanthaster planci* are a normal part of Indo-Pacific coral reefs. They are the cause of the largest known pest-related disturbances on Indo-Pacific reef systems and are regarded as a major management problem (see Moran 1986 and Birkeland and Lucas 1990 for reviews). The starfish eats coral tissue, favoring abundant fast-growing plate and branching corals but also consuming the tissues of rarer, slow-growing massive types that take much longer to replace themselves (Done 1987, 1988). The starfish have periodically killed more than 90 percent of the corals on many reefs throughout the region since the 1960s, returning to individual reefs after about fifteen years. In the first few years following such outbreaks, algal-covered reefscapes are drab compared to their coral-dominated predecessors and successors.

Human-Induced Disturbances

Humanity may be affecting the frequency, intensity, and distribution of many types of disturbance. For any given reef in the Great Barrier Reef, for example, the average interval between cyclone impact and inundation by a flood plume is likely to decrease under all global climate change scenarios (Pittock 1999). The amount of terrestrial sediments re-suspended by those storms and delivered by those floods will likely increase as a result of elevated soil erosion from land used for many decades to centuries for various urban, industrial, water conservation, and rural activities. Humans, through fishing down of keystone predators (Ormond et al. 1988; McClanahan 1995c) or elevating nutrients (Birkeland 1982), may increase the frequency and intensity, and increase the geographic extent of, coral predators such as crown-of-thorns starfish (*Acanthaster planci*) and gastropods (*Drupella* spp.).

Nutrients and Sediments

On the Great Barrier Reef, small-scale experiments have in the past indicated negative effects of inorganic nutrient additions on two major ecological functions of reefs (figure 5.1), namely total primary productivity (Kinsey and Domm 1974) and coral calcification (Kinsey and Davies 1979). More recent controlled exposures to ammonium and phosphate, either individually or together, indicated that the most important primary producers, the epilithic algae, may be nutrient sufficient (Larkum and Koop 1997). In fact, in some areas, factors other than nutrients may be limiting to algal production, such as light (Adey and Goertemiller 1987), inorganic carbon supply, space availability (Williams et al. 2001), or, indirectly, water movement (Larkum and Koop 1997). Nutrients may stimulate the production of algae that could potentially overgrow and kill corals (Tanner 1995; Lapointe et al. 1997). Nonetheless, other disturbances such as grazing and saltation of sediments often compensate for this increased growth (Hatcher and Larkum 1983; McClanahan 1997a; McCook 1999) such that the predicted competitive exclusion does not occur. It appears that many reefs are, therefore, resilient to inorganic-nutrient additions, and this may be governed by a number of circumstances. Thus, reefs generally have substantial nutrient stores in sediments and biota, and external supply may be substantial (Tribble et al. 1994; Polunin 1996). There is better evidence, however, that erosion of reef structure increases with increased nutrients (Risk et al. 1995).

Large-scale nutrient additions to reefs from human activities will generally come from the land, but these additions are often accompanied by other changes in water quality, such as increased concentrations of suspended particulate matter and reduced salinity. Thus, where large-scale degradation has occurred in inshore reef communities, attribution of this degradation to specific factors such as nutrients is uncertain (Kinsey and Davies 1979; Tomascik and Sanders 1987; Tomascik et al. 1997).

Corals are generally considered to be sensitive to sedimentation effects, but primary framework-building corals vary in their ability to withstand sediment deposition on them and some may thus be considered resilient (Rogers 1993). Sediments may change the generic composition of corals but not the total coral cover (McClanahan and Obura 1997) or may arrest the successional development of algae at the turf stage rather than accelerating it toward erect fleshy algae (McClanahan 1997a). Because these studies were undertaken in marine protected areas, where fishing was excluded, it was easier to distinguish river sediment effects from other likely human influences. Whatever the specific mechanism, coral reefs in areas with human development can be expected to display ecological changes. These include the maximum depth of water to which corals grow (Tomascik et al. 1997), coral community composition (Randall and Birkeland 1978; van Katwijk et al. 1993), fish abundance and species composition (Amesbury 1981; Green et al. 1997), and processes such as grazing and sediment turnover (McCook 1996; Green et al. 1997).

Resource Extraction

Fisheries are the most extensive extractive use of living reefs. Fishing reduces the abundance and mean size of target species such as snappers (Lutjanidae), groupers (Serranidae), grunts (Haemulidae), and emperors (Lethrinidae), and it is evident, even on lightly fished reefs, that the decline in biomass may be rapid (Jennings and Polunin 1996a). Since many such fishery-target species are carnivorous, the abundance of whole trophic groups of fishes, such as invertebrate-feeders, is thought to be sensitive to exploitation (figure 5.3a). This swift depletion of target species is the basis for contention that reefs are sensitive to fishing, and indeed the catch per unit effort does decline in a similar fashion (figure 5.3b). However, as in other fishery stocks, decline in biomass is evidently accompanied by increased productivity (Jennings and Lock 1996). The stocks

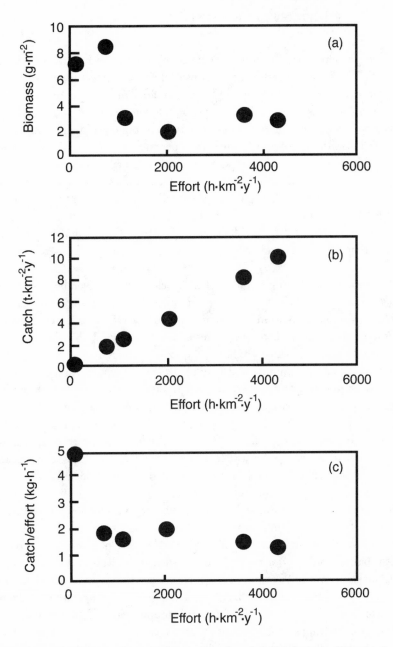

Figure 5.3. Depletion and sustainability of small-scale reef fisheries in Fijian traditional fishing grounds (*qoliqoli*) subject to different levels of fishing pressure: *(a)* decline in biomass of invertebrate-feeding fishes estimated by underwater visual census (UVC) at low levels of fishing effort (Jennings and Polunin 1996a), *(b)* evidence that the catch rate per unit of reef area increases linearly as fishing effort increases (after Jennings and Polunin 1995a), and *(c)* catch and effort data corroborate those from the UVC analysis (after Jennings and Polunin 1995a).

involved, therefore, may be quite resilient to exploitation in productivity terms (figure 5.3c). In spite of earlier impressions to the contrary, reef fishery stocks in the South Pacific are able to support yields that are high on a per unit reef area basis (fresh weights of 10 to 30 tons per square kilometer per year). The sustainability of these high levels of productivity is corroborated by modeling studies (Polunin et al. 1996) and by their own persistence in time (more than ten years), although in interpreting existing fisheries yield data, care should be taken with the units of comparison (Munro 1996). Field studies on Kenyan coral reefs suggest that fish catch, through the use of seine nets that catch small fish, is reduced on a per-area and per-effort basis compared to sites without this gear (McClanahan et al. 1997; McClanahan and Mangi 2001). Modeling studies suggest that catch rates and selection can affect fisheries yields (McClanahan 1995b).

Substantial contributions to reef fisheries yields may, however, be made from adjacent unexploited coral reefs or from other ecosystems such as seagrass, plankton, and sandy-bottom ecosystems (Polunin 1996). Consequently, high fishing pressure may in many such cases be substantially supported by recruitment from distant reefs that are lightly exploited, as may be the case in archipelagoes such as Fiji (Jennings and Polunin 1996a). On many Kenyan fringing reefs, fisher folk fishing within protected lagoons often catch fish that have migrated in from deeper and less-fished reefs. Additionally, many of the coral-associated species are not an important part of the fisheries at high levels of fishing effort. The catch from coral reefs is frequently composed of species of generalists that are only weakly associated with coral reefs, or of species more frequently associated with other ecosystems or food sources, such as plankton, sandy bottom, or seagrass meadows.

Fishing has had profound and often indirect influences on reefs in areas such as the Caribbean and Kenya (figures 5.2 and 5.3, and see below). A number of studies have shown increased abundance and diversity of small-bodied damselfish, parrotfish, and wrasses on heavily fished reefs (Russ and Alcala 1989; McClanahan 1994, 1997c) and losses of species at the highest levels of fishing (McClanahan 1994, 1997c; Ohman et al. 1997). A comparative study of Kenya's fringing with patch reefs of Tanzania found a 50 percent loss of species diversity on the fringing reefs but not the patch reefs (McClanahan 1997c). This may be due to habitat differences. It is more likely, however, that the fringing reef environment has the effect of compressing fisher folk behind the reef, resulting in high population densities (7–14 per square kilometer) in shallow (less than 5 meters deep) reef and seagrass habitats (McClanahan and Kaunda-Arara 1996; McClanahan et al. 1997). The densities of Tanzanian fisher folk are less (2–4 per square kilometer) and reefs are typically separated by deeper water

(McClanahan et al. 1999). Consequently, many of the indirect changes associated with fishing may be dependent on the density of fisher folk and their catch selection, as also suggested by model simulations (McClanahan 1995b), and the densities on Fijian reefs are typically low.

Unsustainable exploitation may occur where fishing is extensive and recruitment overfishing occurs. In the central Pacific, there is as yet little evidence that these high levels of human disturbance have detrimental side effects on the wider ecosystem. Koslow et al. (1988) presented evidence for changes occurring in the fish community of Jamaican reefs as a result of exploitation, but these may be best explained by selectivity of the fishing process, and ecosystem or indirect effects of fishing are not necessarily involved. The expectation that prey fishes, such as those of small size, or those of many nontarget groups, should increase in abundance when piscivorous fishes are removed by fishing has not been corroborated by work focused on some of the larger target species (Jennings and Lock 1996; Jennings and Polunin 1996a).

Interactions between Disturbances

In many cases, disturbances interact to cause ecological change on coral reefs. For example, one prediction of the effect of sedimentation on coral reefs is that soft corals should increase in abundance (De'ath and Fabricius 2000), but in Kenya, this appeared to be the case only in areas with high water movement (McClanahan and Obura 1997). Coral bleaching may also result from a combination of elevated water temperatures, low water movement, and light or ultraviolet radiation (Gleason and Wellington 1993; Brown et al. 1994; Dunne 1994), and each of these factors may modify the level at which bleaching occurs. There are probably cases where more than two disturbance factors interact, and future investigations will need to consider this possibility.

Recovery Rates of Reefs

In coral reefs, recovery following catastrophic disturbance may or may not result in a return to the predisturbance community structure (Hatcher 1984; Done 1992; Knowlton 1992; Hughes 1994; McClanahan and Obura 1995). Whatever the specific details, the rate of recolonization and growth is as much a function of the location of the coral reef as a property of the recolonizing populations (Done et al. 1996). Within large and dense archipelagos arranged along major current systems (such as the Great Barrier Reef) most reefs regularly receive dense aggregations of the larvae released from upstream reefs (Doherty and Williams 1988; Oliver and Willis 1987). In this setting, high degrees of gene flow have been demonstrated in a number of invertebrate taxa (Benzie 1994). There is, nonetheless, enormous interannual variation in larvae supply and recruitment success among patch reefs and across whole regions (Doherty and Williams 1988).

At reefs separated from their neighbors by great distances, unfavorable currents, or both (as in French Polynesia), it seem likely that larvae from other reefs would arrive less reliably. There may be intervals of many years, decades, or even longer, between "good years" for exogenous larval inputs, and they would depend much less on them than on retention of their own reproductive output (Planes et al. 1993).

Corals

The Great Barrier Reef, which is a dense archipelago of individual reefs, would appear to provide optimal conditions for coral settlement, recruitment, and growth following natural disturbance and return to normal conditions (Done 1992). The recovery rate of corals on slopes severely damaged by crown-of-thorns starfish decreased with increasing water depth with rates (expressed as percentage of total substratum) of around about 7 percent, 5 percent, and 2 percent annually at 1 meter, 3 meters, and 6 meters in depth respectively (Done et al. unpublished). After fifteen years, some denuded shallow sites had attained near 100 percent coral cover, while deeper sites were mostly 30 percent or less. By far the greatest contribution at all depths was made by the same fast-growing branching and plating corals that were dominant before the predation event. In Indonesia, the same suite of corals colonized a denuded area at a rate exceeding 12 percent per year (Tomascik et al. 1996). The rapid linear extension rate of individual corals came, however, at the expense of a marked reduction in skeletal density.

The Great Barrier Reef and Indonesian examples contrast markedly with examples from the Caribbean (Connell 1997). For example, in the Dry Tortugas, coral settlement densities were only a fraction of those observed on the Great Barrier Reef (Kojis and Quinn 1994). Combined with a tiny post-settlement survival rate and poor supply and survival rates of coral fragments, negligible recovery was observed. Studies on Florida coral reefs (Patterson et al. 1997) indicate a continued and widespread negative impact on coral cover by a variety of diseases.

Recovery in massive corals on the Great Barrier Reef has been assessed using simulation models (Done 1987, 1988b). These suggest that the currently observed fifteen-year recovery interval may be sustainable only on some reefs. The criterion for a "sustainable disturbance interval" was the time necessary, under conditions of simulated recurrent disturbance, for the coral population to consistently maintain a balanced size-class distribution, including the oldest class. Differences among the reefs were a function of the preexisting structure of the target coral population, the size-specific damage regime, and inter-episode rates of coral recruitment, survivorship, growth, and repair. All of these parameters, as well as outbreak intervals of the starfish, may change under global-climate change scenarios.

Fishes

The common larger reef fishes, including coral trout (Ferreira and Russ 1992), parrotfishes (Choat et al. 1996), and snappers, may only rather rarely exceed twenty-five years of age. Consequently, if recruitment is sufficient, the time it will take for a depleted stock to approach its unexploited biomass should usually be twelve to twenty years. In cases where depletion is less severe, substantial effects of protection from fishing can be expected in shorter periods of time, and have in fact been indicated by work on marine reserves in the Caribbean (Polunin and Roberts 1993), small-scale fishing in the South Pacific (Smith and Dalzell 1993), and a newly created marine park in Kenya (McClanahan and Kaunda-Arara 1996). Exceptions to such recovery rates can be expected where recruitment is limited, as may be the case at remote sites in the Pacific and certain areas in the Caribbean (Roberts 1995). McClanahan (2000) found that the triggerfish, *Balistapus undulatus*, a keystone predator, was still recovering in marine protected areas after twenty years of protection from fishing. Additionally, there may be differences in the recovery rates of different fish families or species (McClanahan and Kaunda-Arara 1996), and recovery may also be dependent on the ecological or structural state of the reef. For instance, the recovery rate of reef sites in a newly created marine park appeared to depend on the abundance of sea urchins (McClanahan 1997b). Sites that had been protected from fishing for over three years showed only minimal recovery over this time, but an experimental sea urchin reduction led to a recovery of corals and fishes (McClanahan 1997b).

Other Organisms and Reef Structure

Most reef ecologists have focused on the recovery of corals and fishes, and so less is known about other groups and their recovery times from disturbances. Recovery is, however, often closely related to the territory or size of the individual organisms or the structure of the reef (figure 5.4). Most of the primary production is achieved by reef algae and the recovery rate of these algae is usually less

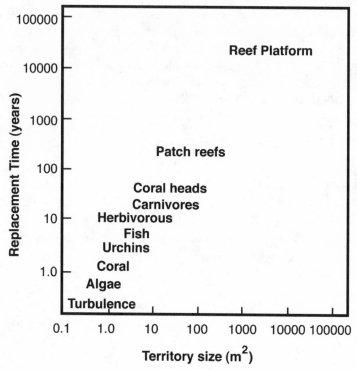

Figure 5.4. Graph showing the replacement time and territory of the major reef components.

than a month for the microscopic filamentous turfs and a few months for the larger erect fleshy and calcareous algae (McClanahan 1997a). The recovery of heterotrophic organisms is usually slower, ranging from years to decades, and the recovery of the reef structures themselves, which include everything from massive coral heads to platform reefs, ranges from decades to millennia. Many of the reef structures we see today have been formed since the last sea-level rise during the past ten thousand years but, in many cases, they were built on reef structures of older origin. Reefs deposit calcium carbonate at a rate of about 1 kilogram per square meter per year, which corresponds to a vertical accretion rate of 0.8 millimeter per year, which is near the global sea level rise of 1.3 millimeters per year over the past one hundred years (Smith 1983). Consequently, reef growth usually keeps pace with the glacially induced sea-level change, but there are some cases, particularly at high latitudes, where reefs can "drown" because they grow too slowly to keep pace with sea-level rise (Grigg 1997). Predictions for the rise in sea level due to global warming are as high as 15 millimeters per year, in which case many reefs could potentially drown if this rate persists over a few centuries (Buddemeier and Smith 1988).

Factors That Contribute to Resilience

Below we describe factors that appear to increase the resilience of coral reefs, as shown by a number of case studies. Resilience is defined here, as in the rest of the book, as the ability of ecosystems to resist lasting change from disturbances (Gunderson et al. 1997) and not the return time of an ecosystem after disturbance (Haydon 1994). In many cases, the quality of science is not sufficient to distinguish between alternative explanations and in other cases it is frequently a number of factors that have either reduced or increased resilience. Consequently, some of the following case studies and explanations should be considered speculative until further studies test the validity of the proposed explanations. The factors proposed to influence resilience should, therefore, be seen as working hypotheses until better tested.

Species Diversity

Evidence is accumulating, particularly from terrestrial plant studies (Naeem et al. 1994; Tilman and Downing 1994; Tilman et al. 1996; Chapin et al. 1998), that "functional diversity" is able to stabilize or buffer ecosystem processes such as resource uptake and productivity. Consequently, species diversity may have the capacity to increase ecosystem resilience by ensuring that there is sufficient informational redundancy to protect against risks associated with environmental disturbances (Naeem 1998). Evidence to support this hypothesis from coral reefs is at present ambiguous. For example, there is little evidence that regions with lower diversity of corals have lower or more variable rates of calcium carbonate deposition, or that areas with lower diversity of algae have lower rates of productivity (Kinsey 1983; Smith 1983). Even though some regions have lower species diversity than others, all maintain similar functional diversity. Functional diversity appears to be maintained within ecosystems even when species diversity is low. The low number of studies and the limited accuracy of measurements of productivity and net calcium carbonate deposition make any conclusions at this point highly speculative, and future work will need to distinguish between functional and species diversity.

Areas with low species diversity, namely the eastern Pacific and the Caribbean, have experienced dramatic changes in their ecology during the past few decades. In the eastern Pacific, the El Niño warming of 1983 led to a series of ecological changes that devastated many reefs in the region. High water temperature caused bleaching and mortality of 70 to 95 percent of corals in Costa Rica, Panama, Columbia, and Ecuador (Glynn et al. 1988; Glynn and Colgan 1992). High temperatures also extirpated populations of a crustacean guard of the dominant coral *Pocillopora* that left the coral vulnerable to the starfish predator, *Acanthaster planci* (Glynn 1987). The final insult to these reefs was a largely unexplained increase in the populations of a sea urchin, *Diadema mexicanum*, which, through its feeding activities, began to erode the reef framework to such an extent that many of the reefs are disappearing (Reaka-Kudla et al. 1996). The 1983 El Niño had a one-in-one-hundred-year return frequency, so perhaps reefs are able to form and recover on this time scale. However, most eastern Pacific reefs are low diversity, seldom form extensive reef flats, and are sparsely distributed.

In the Caribbean, a series of diseases has resulted in a basin-wide reduction in the primary reef-building coral, *Acropora palmata* (Antonius 1981; Gladfelter 1982; Aronson and Precht 1997), and in an important grazer, *Diadema antillarum* (Lessios et al. 1984). Many of the reefs in this region are now dominated by erect algae (Carpenter 1990a; Hughes 1994; Shulman and Robertson 1997; McClanahan and Muthiga 1998), and this, in turn, is suppressing a number of fish species and their grazing rates (McClanahan et al. 2000a, 2001). Although there are few reported studies, it is fair to assume that these reefs have changed not only in species composition (McClanahan et al. 2001) but also in terms of the important ecological processes such as organic (Carpenter 1988) and inorganic production (Sammarco 1980, 1982). In addition, the recovery rate of corals after disturbances in the low-diversity Caribbean has been slower than recovery in the high-diversity Indo-Pacific (Connell 1997). Consequently, there is some cause to believe that species diversity increases the capacity of reefs to tolerate and recover from disturbance, but more research into the mechanisms is required.

Keystone Species and Redundancy

Conventional ideas linking diversity and resilience may or may not be useful in understanding coral reefs. High species diversity may confer a measure of redundancy that maintains ecological processes when individual species decline (McNaughton 1977; King and Pimm 1983; Tilman et al. 1996; Naeem 1998), even though stability of species can be reduced in high diversity systems (May 1977; Haydon 1994). In practice, there may be sufficient biological differences among species such that species are not always fully redundant, interchangeable, or replaceable (Rowan et al. 1997). Disturbances may, over the long term, detrimentally affect specialists more than generalists because specialists are often slow growing, site attached, territorial, and competitive dominants (*K*-selected), while generalists will frequently exhibit the opposite traits (*r*-selected). Disturbances to specialists may result in their replacement by fast-growing and vagile generalists until specialists recover and reoccupy their niche. Differences in species' use of resources may result in only partial filling of the specialist's niche. In other words, it cannot always be assumed that generalists will always replace the role of specialists, nor that one specialist replace the role of another, even if they appear to be fairly similar to each other ecologically. Consequently, the loss of key species, even in high-diversity systems like coral reefs, can result in changes in reef ecology that alter resource use, productivity, or accretion.

Loss and Replacement of Keystone Species

Two cases illustrate the effect of either the loss or replacement of keystone species on the structure and function of reef systems.

SEA URCHIN PREDATOR CASE STUDY

Relationships between sea urchins and their fish predators can greatly affect reef ecology. Sea urchins have been reported in the stomachs of a large number of fish species (Randall 1967), and it might, therefore, be inferred that one sea urchin predator could easily be replaced by another predator. However more careful analyses of sea urchin predators suggest that complete replacement of one predator by another is, in fact, unlikely.

First, many of the species having sea urchin contents in their guts are not actually predators of sea urchins but, rather, scavengers of dead sea urchins (McClanahan 1995c, 1999) (figure 5.5). There are fewer true predators than scavengers, and in many cases one or a few species are doing most of the predation (McClanahan 1995c, 1999, 2000). Of the eight species of predators of the sea urchin *Echinometra mathaei* in Kenya, one species, the red-lined triggerfish (*Balistapus undulatus*), preyed on over 80 percent of experimental urchins. This fish exhibited sophisticated eating and foraging habits compared to wrasse and scavenger predators (McClanahan 1995c). The other predators were a variety of large, often-terminal male wrasses, and a few other invertebrate-feeding fishes (Lethrinidae) that were also common scavengers of sea urchin carcasses. At Mombasa Marine National Park, there was a change in the dominant predator shortly after the cessation of fishing, as the triple-lobed wrasse (*Cheilinus trilobatus*) was replaced by the red-lined triggerfish during the first few years after the park's creation (McClanahan 2000). Observations indicate that the red-lined triggerfish is aggressive toward other predators like the triple-lobed wrasse, and

(a) Belize - Caribbean

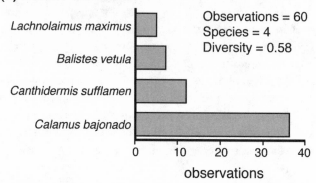

Observations = 60
Species = 4
Diversity = 0.58

(b) Old Kenyan Marine Parks - Indian Ocean

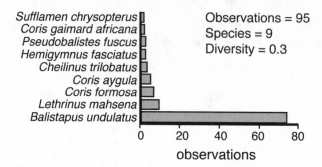

Observations = 95
Species = 9
Diversity = 0.3

(c) Mombasa Marine National Park

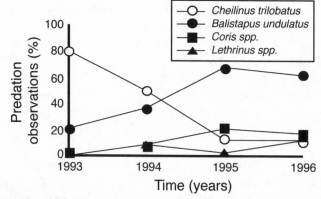

Figure 5.5. Predators of *(a)* the sea urchin *Echinometra viridis* in Glovers Reef Atoll (Belize), *(b) Echinometra mathaei* in Kenyan marine parks (western Indian Ocean), and *(c)* the change in predators in the Mombasa Marine National Park over its early stages of protection from fishing.

as its populations recovered it was able to reduce the frequency of predation of its competitors (figure 5.5c).

The red-lined triggerfish is very uncommon in shallow water outside of East Africa's marine protected areas while the wrasses and scavenger species are relatively common (McClanahan 1994, 1997c, 2000). The aggressive behavior of the triggerfish, which may make it the competitive dominant in the sea urchin predator guild, is a disadvantage in heavily fished reefs because it is often among the first species to take bait in traps or on lines. It may also recover slowly after being eliminated by fishing (McClanahan 2000). The red-lined triggerfish is, therefore, uncommon in a well-developed fishery, while the other predators such as some of the wrasses and scavengers are able to tolerate fishing disturbances and continue to be common in the catch of mature fisheries (McClanahan personal observation). Consequently, many of the fished reefs in East Africa have lost the dominant sea-urchin predator through fishing, and the predator role is largely being assumed by other species of wrasse and scavengers.

Sea urchin densities are typically far greater on fished reefs than in older and fully protected marine parks (McClanahan and Shafir 1990; McClanahan 1998). It appears that the subdominant and more naive predators are unable to maintain sea urchin populations at the low levels found in reefs undisturbed by fishing. An increase in sea urchin populations has a series of ecological effects on the reef. A beneficial effect of the sea urchin increase is a reduced biomass of fleshy algae and possibly increased net production and nitrogen fixation (Carpenter 1988; Williams and Carpenter 1988; McClanahan 1997a). Detrimental affects are increased erosion of the reef substratum (Birkeland 1988; McClanahan 1995b) competition with, and a reduction of, herbivorous and other invertebrate-feeding fishes (Carpenter 1990b; Robertson 1991; McClanahan et al. 1994, 1996), and, at the highest levels of sea urchin abundance, a loss of coral cover (Sammarco 1980; McClanahan and Mutere 1994). Consequently, the accumulation of sea urchins is probably associated with an increase in net organic production but a decrease in inorganic carbon production. Much of this net organic production is not transferred beyond the sea urchin grazer guild, so it does not benefit fisheries (McClanahan 1995b). In fact, there is accumulating experimental and theoretical evidence that sea urchin dominance reduces fisheries production (McClanahan 1995b, 1997b; McClanahan et al. 1994, 1996).

In the Caribbean, the queen triggerfish (*Balistes vetula*) may be an ecological equivalent to the red-lined triggerfish in the western Indian Ocean (although experimental evidence is sparse). At remote, but fished, Glovers Reef Atoll, Belize, this species was the third most common predator of *Echinometra viridis*. The more generalized jolt head porgy (*Calamus bajonado*—Sparidae, figure 5.5) was the dominant predator (McClanahan 1999). This may reflect a similar situation to that found during the early stages of the Mombasa Marine National Park, where a subdominant is the main predator because the dominant has not recovered from the effects of fishing. Further research will be needed to test this hypothesis. Queen triggerfish populations have been reduced to such an extent in the Caribbean that it is being listed as one of the regions threatened species (Hudson and Mace 1996).

LOSS OF A SEA URCHIN

On relatively undisturbed Caribbean reefs, major groups of grazers, particularly sea urchins and herbivorous fishes, contribute substantially to the maintenance of hard-coral cover and recruitment by suppressing algal biomass (Sammarco 1980, 1982; Lewis 1986; Hughes 1994). A certain balance is maintained between the grazer groups by predation and competitive interactions (figure 5.6a). It is likely that the black-spined sea urchin (*Diadema antillarum*) became an important grazer in the Caribbean because both its main fish competitors (including parrotfishes and surgeonfishes) and its predators (including wrasses, triggerfishes, and porgies) were susceptible to fishing pressure. Grazing by an increased abundance of the sea urchin was able to compensate for the loss of fish grazing pressure caused by fishing (figure 5.6b). *D. antillarum* thus came to play a pivotal grazing role in the ecosystem (figure 5.6c).

In 1983–1984, *D. antillarum* declined by about 98 percent across the region, as the result of a pathogen (Lessios 1988a). Experimental removal of *D. antillarum* had earlier shown its role in controlling algal abundance in shallow reef sites (Sammarco 1982). Although there was scope for other herbivores to compensate in grazing (Carpenter 1990b; Robertson 1991; McClanahan et al. 1994, 1996), remaining herbivorous fishes and other species of sea urchin were not able to achieve the grazing levels maintained by *D. antillarum* (Hughes et al. 1987; Lessios 1988b). This is probably because of continuing fishing pressure (Hughes 1994) and perhaps because there was insufficient ecological redundancy between *D. antillarum* and other unfished species (McClanahan 1999). As a result, fleshy frondose algae quickly became dominant (figure 5.6d). Fleshy frondose algae were also promoted by coral mortality as a result of cyclones, coral bleaching (e.g., Williams et al. 2001), and perhaps elevated nutrients (Shulman and Robertson 1997; Lapointe et al. 1997). In this case, therefore, the system was initially resilient to a fundamental shift in structure because macro-algal suppression was maintained by more than one group of grazers. As far as it is possible to tell, resilience was lost when the grazing function became largely reliant on one grazer, which was susceptible to a pathogen. On deeper Caribbean reef areas (e.g., 12–15 meters), the role of urchins in reducing macroalgal overgrowth was small even before the mass mortality. A negative

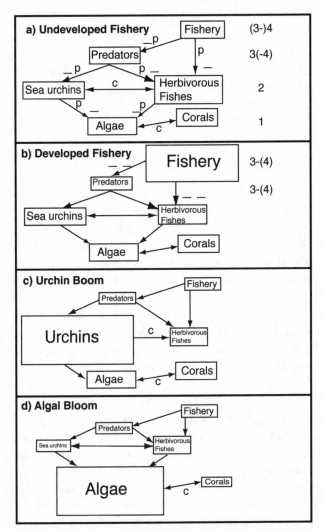

Figure 5.6. Schematization of changes that are thought to have occurred in shallow-water coral reefs of the north coast of Jamaica. Changes in the size of compartments between diagrams suggest changes in the importance of the group or activity (fishery) at each stage, but relative sizes of compartments within diagrams are arbitrary. *(a)* With light fishing, it is postulated that the biomass of predatory and grazing fishes would have been comparatively great, the abundance of sea urchins would have been relatively low and a certain balance would have been maintained between corals and macroalgae; it is possible that this ceased to be the case many decades ago (Jackson 1997). *(b)* The fishery depleted predatory and herbivorous fishes, so reducing predation and competitor pressure, respectively on sea urchins. It is supposed that macroalgae did not increase because sea-urchin grazing was substituted for fish herbivory to some extent. Corals remained abundant. *(c)* Substantial depletion of predators and herbivorous fishes is thought to have led to a sea-urchin boom, which in the early 1980s at least was curtailed by the impact of a microbial pathogen. *(d)* The reduction in total grazing then led to a bloom of macroalgae, which outcompeted the hard corals. The latter were also reduced by hurricanes. The numbers on the right-hand side of *(a)* and *(b)* indicate the trophic levels principally involved. The letters *p* (in *[a]*) and *c* (in *[a]*, *[c]*, and *[d]*) indicate predatory interactions and competition, respectively. The minus signs indicate negative effects of interactions (– = significant, – – = substantial)

correlation between abundance of parrotfishes and surgeonfishes and macroalgal cover (Williams and Polunin 2001) is indicative of these grazers controlling that cover today. In shallow reef areas, it is not clear how a return to a balanced grazing by both urchins and grazing fishes might be accomplished.

Spatial Heterogeneity and Refugia

Coral reefs can be resilient to multiple scales of disturbances (Pandolfi 1996; Connell 1997). One important factor that maintains this resilience is the scattered patchy distribution of reefs throughout tropical ocean basins (UNEP/IUCN 1988). The reefs are thus open to recruitment from other reef sources outside of the disturbed areas. Ocean-wide currents can potentially deliver larvae across hundreds to thousands of kilometers (Roberts 1997), although actual dispersal may be more limited (Cowen et al. 2000). Consequently, the combination of spatial heterogeneity and refugia of reef systems, the temporal heterogeneity of dispersal, and a physically stable but moving transport system of currents ensures the connectivity among reefs that are required for recovery. This is an important aspect of ecological resilience.

MARINE PARKS AS REFUGIA

Heavy fishing is now pervasive throughout the tropics and undoubtedly extends farther offshore than in the past (see chapters in McClanahan et al. 2000b). Pollution is often diffuse and many tropical watersheds are heavily influenced by human activities (Lapointe et al. 1997; Hodgson 1997). Consequently, spatial refuge for many reef organisms and ecosystems is often limited to marine protected areas or deep and remote reefs (McClanahan and Obura 1995). The absence of any systematic differences between effectively protected and unprotected areas in the Caribbean indicates, however, that linkages between fishing and benthic structure are weak in these reefs (Williams and Polunin 2000). However, some theoretical (Bohnsack 1993; Roberts and Polunin 1991, 1993) and mathematical modeling (DeMartini 1993; Man et al. 1995; Holland and Brazee 1996; Sladek Nowlis and Roberts 1997), and field studies (Alcala and Russ 1990; McClanahan and Kaunda-Arara 1996; McClanahan and Mangi 2000) of marine protected areas support the possibility that refuge from human disturbances could increase the total resilience of coral reef fisheries by offering refuge to the larger and reproductively mature individuals and to their required habitat.

Two field investigations of the fisheries adjacent to marine protected areas in the Sumilon Island Park in the Philippines and the Mombasa Marine National Park suggest that parks can improve fisheries yields such that they may partially compensate for the lost fishing area. A seven-year study of the fish catches adjacent to the Mombasa Marine National Park found that, although 65 percent of the fishing area was taken up by the park's establishment in 1991, the total catch decreased only by 35 percent, due to a 110 percent increase in the catch per person of those fishers that remained (figure 5.7). Once the park was reduced from 10 to 6 square kilometers in October 1995, the catch quickly increased and then declined to a level just short of the total catch of the area without protection. The coincidence between beach seine exclusion and the reduction in the refuge make it difficult to distinguish the effect of the park's size and the fisheries management adjacent the park's border. There was, however, evidence that the degree of spillover depends on the adjacent management (McClanahan and Mangi 2000) and the time since the creation of the protected area (Russ and Alcala 1996; Sladek Nowlis and Roberts 1997). In Sumilon Island, however, a similar pattern in fish catch was observed when a small park covering 25 percent of the reefs of the island stopped functioning and fishing began inside the park. The loss of the park resulted in a 54 percent loss in the total yield of fish from the island (Alcala and Russ 1990). These experimental management and modeling studies combined suggest that spillover from refuge such as marine protected areas have a partial ability to compensate for lost fishing grounds and decrease variability in fish catches.

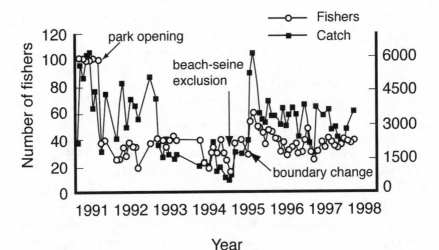

Figure 5.7. Time-series data of the number of fishers and their individual and total catch for a period in which a marine park was created, the size of the park was reduced, and beach-seine fishing was excluded.

CONNECTIVITY

Over the long run, isolation can have devastating consequences to ecosystems. While connectivity can aid maintenance of ecological resilience, it can also expose the ecosystem to invasive species and thereby reduce, or at least challenge, the ecosystem's resilience (Simberloff 1991). This is true when the species assemblage is exposed to novel disturbances such as a new pathogen or predator of a keystone species as described above for *Diadema antillarum*. Novelty can challenge ecosystems' resilience, but it is also arguable that over the long-term, there is self-organization around novel events or organisms that incorporates novelty into the total biological diversity. This new diversity, in turn, becomes a further part of the ecosystem's resilience. Consequently, the temporal scale of observation can result in different conclusions concerning connectivity and resilience.

The complementary alternative to the above novelty-resilience hypothesis, but still relevant to connectivity, is the *stress model*. This model makes a series of predictions about the structure of species assemblages based on the idea that smaller and more isolated water basins experience more physicochemical environmental variability than do larger and open environments (Odum 1985; Bertness and Hacker 1994; Kareiva and Bertness 1997). The stress model predicts that small and isolated ocean basins will have (1) fewer, (2) smaller-bodied, and (3) less-ornate species that largely feed low in the trophic web and which are usually detritivores and herbivores (Vermeij 1978; Bertness 1981; Odum 1985; McClanahan 1992; Rapport and Whitford 1999). These predictions are, however, based more on population dynamic considerations (Pimm 1982) and less on an adaptive system or species accumulation model (Odum 1983; Kauffman 1993). Fluctuating environments will largely and by chance have fewer species able to tolerate the environment, and those that can tolerate it are likely to feed on the most abundant food and have fugitive strategies. Consequently, large, ornate, and oxygen-demanding bodies are likely to do poorly in stressed environments.

An example in support of this stress theory is the assemblage structure of gastropods in shallow waters of different ocean basins (Vermeij 1978; Stanley 1986; McClanahan 1992). Gastropods are a good assemblage with which to test some of these hypotheses because they are largely adapted to hard bottom aquatic environments, they are relatively easy to see and identify, are noncolonial, eat a variety of foods, and leave a fossil record. Based on searches of shallow-water environments in the Caribbean (Belize and the Florida Keys) and the Indian Ocean (Kenya and Madagascar), there is good support for these hypotheses (figure 5.8). In addition, a similar pattern was seen when comparing a bay (Florida Bay) in the Caribbean with a more open environment (Hawk Channel, figure 5.9, McClana-

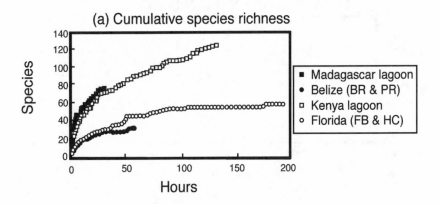

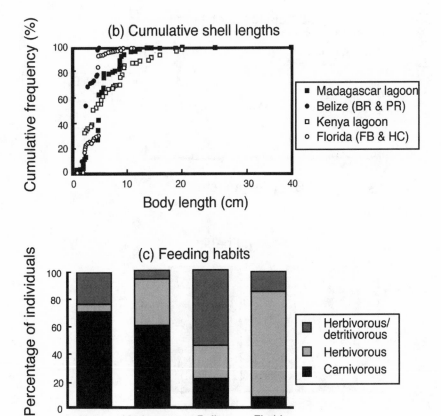

Figure 5.8. Comparison of the prosobranch snail populations in two regions of the western Indian Ocean and Caribbean (McClanahan unpublished data): *(a)* cumulative species observed as a function of cumulative search time, *(b)* cumulative frequency of shell lengths, and *(c)* feeding habits.

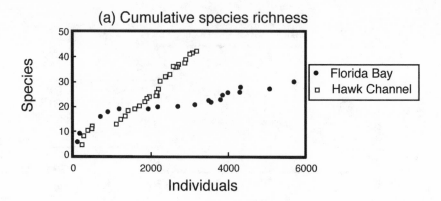

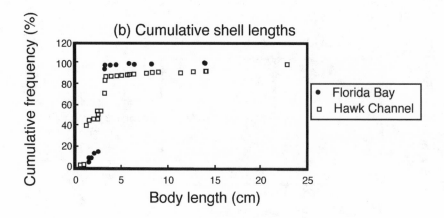

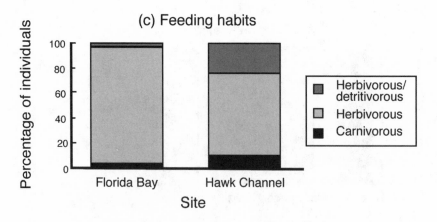

Figure 5.9. Comparison of the prosobranch snail assemblages on the oceanic (Hawk Channel) and bay sides of the Florida Keys (data from McClanahan 1992): *(a)* cumulative species observed as a function of cumulative search time, *(b)* cumulative frequency of shell lengths, and *(c)* feeding habits.

han 1992). Consequently, isolation should have the effect of increasing the variation in the physical environment that, in turn, should affect ecological and evolutionary organization of the species assemblages. This ecological organization can be independent of the level of connectivity but more a measure of species' tolerance to environmental fluctuations. This model predicts the opposite of an ecosystem adaptive-learning model. It predicts that smaller and more fluctuating environments will be more resilient to disturbances because the species that persist in these environments are already adapted to fluctuations and are less likely to be perturbed by new or additional fluctuations.

The alternative hypotheses are more compatible when one considers that the organization of species assemblages under different levels of environmental stress is often a process that works more on ecological than on evolutionary time. In fact, many of the species in stressed environments are ancestral species or species that were formed prior to the formation of the organisms that now inhabit less environmentally stressed surroundings (Vermeij 1978). For instance, the evolutionary origin of most of the marine snail species found in the Florida Bay is on the order of many millions of years, while the bay itself was formed by the recent sea level rise and is, therefore, less than ten thousand years old. These ancestral species have not persisted in much of their original range because, in theory, they have been replaced by more recent evolutionary forms and presumably find refuge from detrimental species interactions in stressed or low-productivity environments that are uninhabitable by their potential competitors. In essence, these ancestral species find refuge from biological forces like predation that are more intense in the less environmentally stressed surroundings (Vermeij 1978; McClanahan 1992). Consequently, species assemblages in stressed environments have not been an active part of the main evolutionary process but are relicts of and peripheral to it.

Ecological stress and information accumulation, or evolutionary and ecological organization, are counteracting forces. Environmental stress works to reduce species diversity and redundancy to those species most able to tolerate large environmental fluctuations but not additional species of potential competitors and predators. Ecological and evolutionary organization often work to increase species diversity and redundancy as species divide up resources more finely over evolutionary and ecological time. Despite the long-held view that closely related species cannot coexist, more recent models suggest there is more tolerance of species overlap than previously recognized (Abrams 1996).

Synergistic Stressors

The number and extent of environmental stresses can influence resilience of coral reefs. This occurs because stresses can act in multiplicative rather than additive ways, or because the sum of two stresses can exceed a threshold where a single stress would not. Consequently, two or more stresses or disturbances working independently may have a much smaller effect than the two factors working together. The opposite or inhibitory interactions or synergies can also occur when one factor would have an effect except that a second factor is nullifying this influence. There are numerous examples of these types of interactions, and we will briefly describe some examples relevant to coral reef conservation and management.

Reef ecologists believe that the largest human influences on coral reefs, manageable on a local scale, are fishing, pollution, and sedimentation (Ginsburg 1994). Because these influences are common throughout the tropics, it has been difficult to distinguish their independent influences. In order to study the effects of nutrification alone, it may be necessary to study its effects in marine protected areas that exclude fishing or in areas without human influences. These factors are frequently associated with human population densities, and few governments will establish marine protected areas in polluted waters or allow marine scientists to artificially pollute protected areas. However, there are studies that suggest that the response of coral reefs to these factors combined may be different from them in isolation (e.g., Umar et al. 1998; McCook 1996, 1999; Miller and Hay 1996).

The Malindi-Watamu Marine National Park in Malindi, Kenya, is an example of a protected area that has been nutrified over the last two decades. The park has been receiving increasing sediment and nutrients from the Sabaki River over the last fifty years associated with an increase in land uses that promote soil loss (Dunne 1979; Finn 1983). However, this river influence appears to have only minor influence on the corals (McClanahan and Obura 1997). The total cover of hard coral and algae benthic cover remained nearly the same over a seven-year period despite periods of three months or more per year when waters were brown and turbid. The largest change over the study period was a shift in the species composition of the corals such that those species more tolerant of the increased nutrient and sediment conditions increased while less-tolerant species decreased. This minor shift in species composition was unexpected. Many reef ecologists believe that, depending on levels of herbivory and other disturbances to algae, nutrified reefs will be dominated by fast-growing reef algae that will eventually overgrow and kill corals (Littler et al. 1991; Bell 1992; Delgado and Lapointe 1994; Lapointe et al. 1997).

The unexpected response was probably attributable to two factors: sediments from the river discharge, and a lack of fishing. The sediments acted as a physical disturbance that retarded ecological succession to an early-successional turf algae stage rather than a carrier of nutrients that accelerated it toward fleshy algae dominance (McClanahan 1997a). Lack of fishing meant that herbivorous fishes were abundant in this reef, and the persistence of these consumers may have kept fleshy algal abundance in spite of any stimulation of growth by nutrients. This interpretation is consistent with the views of Szmant (1997) and McCook (1999), which suggest that the maintenance of fishes and their feeding responses as well as the physical structure of the reef may be important factors that can inhibit algal overgrowth of coral reefs. These studies suggest that it is often the combined effect of the loss of reef consumers, the reef structure, and nutrification that cause drastic changes in reef ecology. Consequently, if reefs are being nutrified, it may prove useful to restrict or eliminate fishing to improve the chances of maintaining coral phase ecology.

A second example of increased resilience was shown during a sea urchin reduction experiment in the Mombasa Marine National Park, which was created in 1991 (McClanahan et al. 1994, 1996). Sea urchins are so abundant in many Kenyan reefs that they can be considered pests that require active management. Consequently, this study reduced the abundance of sea urchins by about 85 percent in the park, as well as at two fished reefs. In the fished reefs (low fish herbivory) where sea urchins were reduced, fleshy algae and seagrasses colonized and reduced living coral by about 35 percent over the one-year study. In contrast, the increase in fleshy algae in the Mombasa Park (high fish herbivory) was smaller and resulted in little change in the coral cover. The reason for these differences is, again, attributable to the greater abundance of surgeonfish and parrotfish in the park, which increased after the sea urchin reduction and kept fleshy algae from increasing and smothering coral. In unprotected reefs, there was a moderate increase in small-bodied damselfish and juvenile parrotfish, but this increase appeared insufficient to keep fleshy algae from dominating the site. Consequently, the greater functional diversity of herbivores in the protected areas acted to inhibit algal overgrowth of the reef and maintain its ecological stability. In general, unfished reefs have greater diversity on the small scale, more ecological redundancy, and, therefore, are more likely to buffer the effects of species deletions or reductions. Consequently, in order to maintain resilience, it behooves managers of coral reefs to maintain species of functional diversity and reduce the number of simultaneous disturbances or stresses.

Human Institutions and Coral Reef Resilience

Humans, through their cultural institutions, have the ability to reduce or increase the resilience of ecosystems depending on various cultural and environmental conditions and interactions. With the exception of the recent rise of tourism and marine protected areas, the human relationship with coral reefs has been that of hunter-gatherer, even when exploited by otherwise agrarian or industrialized humans. Below we will describe, from an historical perspective, some case studies of the development and decay of human cultural institutions associated with coral reef fisheries and discuss their relevance to ecological resilience.

The Economic Evolution of Watersheds

The history of coral reefs adjacent to high islands and continents is often closely associated with the economic development of watersheds. On high islands and continents, coral reefs and their fisheries are often just one of a number of domestic and wild ecosystems that humans will utilize. In the absence of strong cultural taboos, human populations that have a high reliance on agriculture will partition their effort among ecosystems in such a way that their labor is best rewarded with resources in the short term. In many tropical areas, rainfall or other environmental factors required by crops are seasonal, with either one or two planting, weeding, and harvesting cycles per year. When human population densities are low and human labor limits the production of crops, it is common to have people of many ages and both sexes attending to crops during these critical periods. Consequently, at low levels of domestication of the watershed, fishing is often a seasonal activity largely undertaken by men, when their labor is not required for crop production or hunting on land.

As human population densities increase, agricultural labor becomes less limiting, greater partitioning of labor occurs among the various production systems, and a guild of full-time fishers develops. Fishing becomes less a subsistence activity and more a commercial activity where fishers will attempt to catch far beyond their household requirements. As human populations increase, fishing can become one of the few ways to meet deficits and earn extra money to support impoverished people. In such cases, fishes in nearshore waters may be quickly depleted, and fishers adopt boats, technologies, and patterns of movement that will allow them to exploit fisheries farther offshore. Many fishing communities now occupy and exploit land and sea areas with insufficient productivity to fully support them. They, therefore, often emigrate to expand their access to resources.

The development and cultural adoption of fossil fuels and associated technologies greatly increased production and reduced the human labor requirement. It led to further specialization and accelerated aggregation of people into towns and cities. Fishing grounds and technologies expanded, but in probably the majority of cases nearshore fisheries were already fully exploited, and the high capital and costly technologies caused many fisheries to collapse or to expand to offshore waters (Berkes 1987). The pressures on fished populations became governed more by the complex interplay of market demands, supply, subsidies, and the cost of fishing, and less by the direct interactions between local needs, skill, and effort (Thomson 1980). Costly and high-technology fisheries were not necessarily the most beneficial for local communities (Kamukuru 1992).

Associated with the development and expansion of resource use were common patterns of waste production. Sediments and nutrient run-off increased along a gradient from indigenous forest to plantation forest to shifting agriculture to permanent agriculture to pasture (Dunne 1979; Young 1996). Run-off from large tropical cities and towns was seldom treated and further added to the nutrification of the nearshore and coral-reef environments. This run-off in combination with fishing added to the deterioration of the oligotrophic coral-reef environment.

Low-Lying Atolls and Islands

Low-lying islands and atolls can differ from high islands and continents because the lower capacity for agriculture on these islands makes dependence on seafood and trade more important. Consequently, the capacity to maintain human populations on these islands is more limited, and the maintenance of humans may require that they (1) develop trading partners or tourism, (2) rely on offshore fisheries, or (3) develop a fugitive strategy of exploiting resources and moving on. In fact, these three economic strategies and their various combinations are common to many of the low-lying islands of the tropics. For example, inhabitants of the low-lying Maldive Islands off of southern India have historically relied almost entirely on the catch of pelagic tuna and until recently did not fish on their reefs. Recently, outsiders interested in exporting live fish to other Asian countries have developed reef-based fisheries there.

Fisheries Management and Its Contribution to Ecological Resilience

In order to adapt to the changing conditions of the human needs/resource abundance ratio, a variety of fisheries management policies and actions have developed (Berkes 1985; Smith 1988; McGoodwin 1990; Smith and Berkes 1991; Berkes et al. 1995). In the early stages of the economic evolution, this largely included ways to more fully exploit resources, with little efforts toward conservation (Smith 1988). Nonetheless, there is evidence for conservation of resources among small-scale fishers and hunter-gatherer, agrarian, and industrialized cultures that could potentially result in conservation and increased resilience of the human fisheries resource (Johannes 1981; McGoodwin 1990; Smith and Berkes 1991; Zerner 1994; Ruddle 1996; McClanahan et al. 1997). Regardless of the state of the culture, conservation of resources is largely achieved through restrictions on (1) gear, (2) access to the resource or the limitation of effort, and through use of (3) time limits, (4) size restrictions, and (5) spatial restrictions such as sacred, dangerous, or marine protected areas. Below, we will discuss how each of these restrictions has developed in different cultural conditions and their effect on conservation and ecological resilience of coral reefs.

SMALL-SCALE FISHING SOCIETIES: MIJIKENDA OF EAST AFRICA

The Mijikenda are an association of largely agrarian Bantu people that have lived as a distinct ethnic group along the East African coast for approximately four hundred years (Spear 1978). Their oldest settlement is located in the forest watershed of the East African coastal mountain range, but their second oldest known settlement is situated just behind a coral-sand beach adjacent the East African fringing reef (Spear 1978). These people have developed a number of cultural institutions that have parallels with modern fisheries and resource management. These institutions are, however, closely linked to the spiritual view of these people and not explicitly or, perhaps, consciously developed in order to manage fisheries or natural resources (McClanahan et al. 1997; Glaesel 1997).

Fish landing sites in this area are now largely managed by a group of elders, who include two *kaya* elders and two fisheries cooperative leaders. *Kaya* elders are traditional leaders who inherit their positions from their fathers and take this leadership role after the death of their father. Fisheries cooperative leaders are a more modern institution, and these leaders are voted into their positions by the fishers that utilize the landing site. *Kaya* elders teach, oversee, and perform traditional cultural institutions. Interviews with the kaya elders indicate that a number of them believe that their ancestors placed restrictions on the gear used to catch fish, the size of fish caught, access to resources by fishers, and times as well as the places when fishing is allowed (McClanahan et al. 1997). For example, if fishers from outside the community would like to fish in the waters adjacent the landing site, they are expected to pay a fee, called *ubani*, to the elder and this is seen as a communal gift to the fishing community. If the fishers are not wanted for reasons of their gear use or other matters, then the elders can refuse to accept ubani and the foreign fishers are expected to leave and not fish in "their waters." This institution suggests a degree of territorial control of fishing grounds.

Ubani is traditionally used for communal activities such as purchasing goods—a goat or chicken, rice, and sweets—or for the annual sacrifices that these communities perform. During sacrificial times no one is allowed to fish. Additionally, most of these landing sites have a series of areas where these sacrifices are performed, called *kaya* or *mzimu*. *Mzimu* are unusual features such as large coral heads, springs, and caves on the beach or at sea that are believed to be inhabited by spirits that can influence fish catches, births, marriages, and other important events in these peoples' lives. *Mzimu* were identified by ancestors during their dreams, and their locations were passed down through the generations. Traditionally, *mzimu* were considered dangerous locations, because of the spirits, and sites should be visited only when making sacrifices. Consequently, many of these areas were unfished until recently and might be viewed as traditional marine protected areas despite the fact that this was never their stated purpose.

A recent study examined the utility of these institutional practices in sustaining resources by comparing the traditional knowledge, fish catches, and coral reef ecology adjacent to landing sites maintaining different levels of adherence to these traditions (McClanahan et al. 1997). It was discovered that fish catches adjacent to two landing sites with the most-strict adherence to traditional practices had 40 percent higher catches per person than those that did not, but the ecological condition of the reefs was similar and indicated a high level of ecological degradation. One reason for this is that the more tradition-adhering communities did not adhere to all traditions (for example they fished in *mzimu* and young fishers used spear guns, which are not condoned by some *kaya* elders) and they were unable to exclude foreign fishers or those from adjacent landing sites. This was largely because the national government institutions, including police and government fisheries officers, did not allow them to use physical force as part of their enforcement program. These communities were, therefore, limited to passive means of enforcement that included discussions with foreign fishers, not buying or allowing foreign fishers to sell their fish at the landing site, and also pushing their boats onto dry land. Consequently, it was common for foreign fishers to use nontraditional gear such as pull-seine nets in the waters of the tradition-adhering communities. This created a great deal of animosity between the tradition-adhering communities with foreign fishers and their national governments, and may be one of the reasons for the poor ecological state of their reefs.

SMALL-SCALE FISHING SOCIETIES: FIJI AND THE PACIFIC

In the Pacific and Southeast Asia, reef and lagoon resources and areas were widely subject to traditional use rights, and many of these controls persist to this day (Ruddle et al. 1992). In many parts of Indonesia, natural resource use may generally be subject to traditional laws (called *adat*), and management systems such as those in parts of the Moluccas (*sasi*) help to constitute a form of coastal territoriality. In Fiji, traditional fishing grounds are known as *qoliqoli*. Restraint of the level of exploitation by such practices, which are essentially a form of ownership, is nevertheless strongly contested for many areas in Southeast Asia and Melanesia (Polunin 1984). Carrier (1996) specifically argued, for a site in northern Papua New Guinea with a sophisticated pattern of marine tenure, that resource depletion was unlikely to be recognized. Whatever the origins, there is overwhelming evidence that this customary tenure tended and tends to be opportunistic (Polunin 1984); it can scarcely alone restrain current resource use, where a resource suddenly gains in value, or exploitation steadily increases through human population growth (Tomascik et al. 1997).

A forceful case for the development of customary reef tenure having been accompanied by conservation awareness has been made for small islands in the South Pacific, where marine resources have traditionally been an essential source of human sustenance (Johannes 1978). Ruddle et al. (1992) argued that in the South Pacific, where customary marine tenure exists, there are plentiful examples of their resilience to outside pressures. But, for Southeast Asia, there is a lack of critical analysis and where semi-quantitative analysis has been made the conclusion has been that the tenure systems lacked resilience (Jennings and Polunin 1996b).

The Fijian *qoliqoli* appear to vary substantially in their management aptitude; specific regulations are only variably enforced such that their performance in the face of growing resource pressure is unpredictable (Cooke et al. 2000). Smaller Pacific islands are currently the focus for developing sustainable resource use (Dalzell et al. 1996), and, despite the variable performance of existing tenure systems, many will persist and may even be further strengthened by national government constitutions. Consequently, any future management regime will have to cooperate with indigenous tenure even if the evidence for its management resilience, and in particular management aptitude, to internal or external pressures remains weak. If such resilience and aptitude could be demonstrated or strengthened, it could be employed, in partnership with central-government authorities, to maintain resource persistence, or indicate where it is most likely to be achieved through co-management projects.

National Governments

In the global context, national governments are middle-level institutions with both upward and downward responsibilities and opportunities for husbanding coral reefs. Upward, national governments often enter into regional and global agreements, each designed to improve overall human welfare; these include agreements on trade, the environment, climate change, sustainable development, and biodiversity. Downward, their most pressing domestic demands will be to meet the demands and aspirations of a population, most of which will care very little for international agreements. For example, the East African and Fijian examples above have shown that sustainable use of coral reefs is often rooted in tradition, but traditions can be weakened by diminishing resource/people ratios combined with conflicting policies of national governments. Thus it is important that detailed specific description and analysis of the cultural and socioeconomic dimensions of resource use and coral reef ecology is suitably communicated to politicians, policy makers, and resource managers. Works of this type (Birkeland 1997; Hodgson 1997; McClanahan et al. 1997, 2000) may only have an influence when the executive summaries of key findings are made easily accessible to the aides of the responsible government ministers, or to the ministers themselves. If government will exists, such information transfer from credible scientists should be critical for seeking sound decisions and the management needed to produce ecologically and socially beneficial outcomes.

Commitment by national governments to international agreements can influence the national debate over the best ways to implement or develop domestic economic and environmental policies. Different sectors of society have different perspectives on the freeing up of world trade: "The key challenge is to limit the current tendency toward greater use of trade measures for environmental purposes in order to avoid undue erosion of the global trading system" (Drake-Brockman and Anderson 1994). In other words, they are suggesting that bringing environmental criteria to bear on international trade agreements tends to erode the global trading system and hence its capacity to pay for environmental protection. Opponents of current arrangements for world trade, on the other hand, suggest that it is the burgeoning global trading system itself that is ultimately the greatest driver of environmental degradation.

Zarsky and Hunter (1998) suggest that national governments can catalyze good environmental outcomes through use of policy and incentive schemes. They note that the national government usually has the role of regulator and enforcer as well as that of financier of public goods and operator of public utilities. For East Asia, Afseh et al. (1996) suggest that the model of environmental management should be based on the concept of multiple agents and multiple incentives. The three key agents that should interact in setting social norms for environmental management are government, markets, and communities. Government can act directly to influence the other sectors, or it can achieve social goals by indirect action; for example, it can design policies and build capacities that enable communities to play a role in regulating industry and markets (and communities) to play a role in the provision of public goods. In this "triangular model," governments can influence the other two sectors in raising environmental performance, thereby overcoming the three universal obstacles of fiscal constraint, lack of regulatory capacity, and political will.

Depending on whether resources are being under or overused relative to sustainable levels, governments can create incentives or disincentives for markets and communities to regulate resource use around sustainable levels. This is a commonly stated goal of national government practice in fisheries and park management in developed countries. However, its implementation depends on having good information and models on the status and trends of fisheries and having the appropriate citizenry and political will in the face of uncertainty and—often—communities with more pressing short-term personal economic goals.

When communities come to accept the need to implement other ecological measures, such as conservation of biodiversity and protection of indirect ecological functions and services, the need for monitoring and ecological understanding becomes even larger. Consequently, the lack of reliable information for managers and users, their lack of confidence in the information, or the inability of communities to change in the face of this information are all obstacles that can cause sustainable resource management to fail. Management can become passive and rely on environmental and economic disasters as the key form of negative feedback unless there is a knowledgeable and adaptable citizenry (Rosenau 1997; Thompson and Trisoglio 1997). The role of scientists to share their unique knowledge to influence public policy to be guided by knowledge and principles of precaution can not be underestimated (Haas 1997).

National governments are in a position to create international and nationwide systems of marine protected areas with potential to be effective contributors to biodiversity conservation and fisheries support. For example, the United States has its National Marine Sanctuary Program and Australia its National System of Marine Protected Areas. In both countries, the protected areas are "multiple use protected areas" with or without strictly protected exclusion zones, a decision that exposes the viability of the areas somewhat to the uncertainties and risks of fisheries management. Included within these multiple-use protected areas are the Florida Keys Marine Sanctuary (United States) and the Great Barrier Reef World Heritage Area and Marine Park (Australia).

Australian governments from both sides of politics have provided strong institutional support for the management of the 380,000-square-kilometer Great Barrier Reef World Heritage Area as a multiple-use marine protected area. Until 1998, the management of the Marine Park and World Heritage Area was funded by a government appropriation. Now, funds for the management of the area, and for research and monitoring underpinning management, are generated by economic activity in the area—specifically, a few dollars per person per day "environmental management charge" on tourists using a commercial carrier to visit the Area. A 1975 Act of the federal parliament created the Great Barrier Reef Marine Park and the Great Barrier Reef Marine Park Authority, excluded mining and mineral exploration from the park, and prescribed areas ("zones") in which different types and levels of activity are permitted and excluded.

There are many opportunities for community input into zoning and management of the area both through regional management advisory committees and through public review of any proposed changes in regulations. The Authority, an arm of the Australian government, has legally binding agreements with the state of Queensland to jointly develop and implement policies and regulations in relation to people's use of the park and the conduct of fisheries. The fisheries regulations abide by principles of ecologically sustainable development, which it defines as "development (1) carried out in a way that maintains biodiversity and the ecological processes on which fisheries resources depend, and (2) that maintains and improves the total quality of present and future life" (Anonymous 1994).

There are major difficulties in transforming these fine sentiments into effective action. Clearly, for the individual fisher, sustaining poorly comprehended ecosystem processes and biodiversity does not carry the same weight as feeding their family or sustaining their fishing boat or next year's mortgage payments. For peak representative groups of commercial fishers, the issues are how many licenses and how much total catch should be allowed in the different types of fisheries (e.g., line and trawl) to maintain the quality of life and income of the fishers, as well as how to develop more environmentally friendly technologies and determine sustainable catch rates for target and by-catch species. These groups collect substantial levies from their members that are matched by government and used for wide-ranging research. For example, there are major ongoing research and monitoring efforts into coral reef fish and sustainable fisheries (Mapstone et al. 1996), monitoring and ecological studies of water quality (Furnas et al. 1997), reef health (Sweatman 1997), ecological values (Done 1995; Done and Reichelt 1998), and connectivity of representative protected areas (Done et al. 1996). Done and Reichelt (1998) have advocated the use of biodiversity, life history information, and oceanographic and risk analysis in the selection and management marine protected areas (Done et al. 1996).

Concluding Remarks

Most coral reef resources have a number of organizations interested in the welfare of this ecosystem. Each level in this potential management hierarchy may have slightly different desires, local communities being most interested in food resources, national governments in tourism and shoreline protection, and international governments in biodiversity or global element cycles. Consequently, there is a potential for conflict in the types of management that each organization will want as well as the information that they will need for determining the status of their desired objectives. One of the greatest challenges to management, therefore, is to achieve greater synergy among the human organizations in this hierarchy. This has remained difficult because of the diverse cultures of the different organizations involved and the conflict arising from the desire to maintain decision power and resource control by different sectors in society.

Conflicting desires are not easily overcome unless a constructive dialogue is created among the organizations, each organization stating their desires in order that areas of overlap and conflict are identified and mutually acceptable solutions derived from these discussions. Trust is also an additional element critical to negotiations and the political process. All of these suggestions suggest that trustworthy information is important in determining the status of the resources and the views and desires of the human organizations. Consequently, it is important that the organizations trust the quality of the information. This can be achieved if representative members of each level of the human hierarchy are involved in either the collection or confirmation of this information or that all human organizations approve of the group mandated to collect information. Scientists, therefore, have the potential to play a key role in this process because of their shared and, hopefully, objective methodology. Co-management is difficult to achieve because of conflicting desires and because only rarely does any one level in the hierarchy have sufficient resources to research, monitor, and protect the resources unless there are some strong economic incentives to cooperate. Thus, information, key players, connectivity, and cooperative synergy are likely to be important elements, not only in the resilience of natural ecosystems, but also within human organizations.

Literature Cited

Abrams, P. A. 1996. Limits to the similarity of competitors under hierarchical lottery competition. *American Naturalist* 148:211–219.

Adey, W., and T. Goertemiller. 1987. Coral reef algal turfs: Master producers in nutrient poor seas. *Phycologia* 26:374–386.

Afseh, S., B. Laplante, and D. Wheeler. 1996. Controlling industrial pollution: A new paradigm. Policy research working paper no. 1672, World Bank. Available online at http://www.worldbank.org/nipr/work_paper/1672/index.htm.

Alcala, A. C., and G. R. Russ. 1990. A direct test of the effects of protective management on abundance and yield of tropical marine resources. *Journal Consieul Internationale du la Exploration de Mer* 46:40–47.

Amesbury, S. 1981. Effects of turbidity on shallow water reef fish assemblages in Truk, Eastern Caroline Islands. *Proceedings of the 4th International Coral Reef Congress* 6:491–496.

Anonymous. 1994. *Queensland fisheries*. Queensland, Aus.: Queensland Government Printing Office.

Antonius, A. 1981. The "band" diseases in coral reefs. *Proceedings of the 4th Coral Reef Symposium* 2:6–14.

———. 1985. Coral diseases in the Indo-Pacific: A first record. *PSZNI: Marine Ecology* 6:197–218.

Aronson, R. B., and W. F. Precht. 1997. Stasis, biological disturbance, and community structure of a Holocene coral reef. *Paleobiology* 23:336–346.

Bell, P. R. F. 1992. Eutrophication and coral reefs—Some examples in the Great Barrier Reef Lagoon. *Water Research* 26:553–568.

Benzie, J. A. H. 1994. Patterns of genetic variation in the Great Barrier Reef. Pp. 67–79 in *Genetics and evolution of aquatic organisms*, edited by A. S. Beaumont. London: Chapman and Hall.

Berkes, F. 1985. Fishermen and the "Tragedy of the Commons." *Environmental Conservation* 12:199–206.

———. 1987. The common property resource problem and the fisheries of Barbados and Jamaica. *Enviromental Management* 11:225–235.

Berkes, F., C. Folke, and M. Gadgil. 1995. Traditional ecological knowledge, biodiversity, resilience and sustainability. Pp. 281–299 in *Biodiversity conservation: Problems and policies*, edited by C. A. Perrings, K.-G. Mäler, C. Folke, C. S. Holling, and B.-O. Jansson. Dordrecht, Netherlands: Kluwer Academic Publishers.

Bertness, M. D. 1981. Predation, physical stress, and the organization of a tropical rocky intertidal hermit crab community. *Ecology* 62:411–425.

Bertness, M. D., and S. D. Hacker. 1994. Physical stress and positive associations among marsh plants. *American Naturalist* 144:363–372.

Birkeland, C. 1982. Terrestrial runoff as a cause of outbreaks of *Acanthaster planci* (Echinodermata: Asteroidea). *Marine Biology* 69:175–185.

———. 1988. The influence of echinoderms on coral-reef communities. *Echinoderm Studies* 3:1–79.

———. 1997. Symbiosis, fisheries and economics development on coral reefs. *Trends in Ecology and Evolution* 12:364.

Birkeland, C. E., and J. S. Lucas. 1990. *Acanthaster planci*: Major management problem of coral reefs. Baca Raton, Fla.: CRC Press.

Bohnsack, J. A. 1993. Marine reserves: They enhance fisheries, reduce conflicts, and protect resources. *Oceanus* 36:63–71.

Borowitzka, M. A. 1972. Intertidal algal species diversity and the effect of pollution. *Australian Journal of Marine and Freshwater Research* 23:73–84.

Brown, B. E. 1997. Coral bleaching: Causes and consequences. *Coral Reefs* 16:S129–S138.

Brown, B. E., R. P. Dunne, T. P. Scoffin, and M. D. A. LeTissier. 1994. Solar damage in intertidal corals. *Marine Ecology Progress Series* 105:219–230.

Budd, A. F., K. G. Johnson, and T. A. Stemann. 1996. Plio-Pleistocene turnover and extinctions in the Caribbean reef-coral fauna. Pp. 168–204 in *Evolution and environment in tropical America*, edited by J. B. C. Jackson, A. F. Budd, and A. G. Coates. Chicago: University of Chicago Press.

Buddemeier, R. W., and S. V. Smith. 1988. Coral reef growth in an era of rapidly rising sea level: Predictions and suggestions for long-term research. *Coral Reefs* 7:51–56.

Carpenter, R. C. 1988. Mass-mortality of a Caribbean sea urchin: Immediate effects on community metabolism and other herbivores. *Proceedings of the National Academy of Sciences* 85:511–514.

———. 1990a. Mass mortality of *Diadema antillarum* I. Long-term effects on sea urchin population-dynamics and coral reef algal communities. *Marine Biology* 104:67–77.

———. 1990b. Mass mortality of *Diadema antillarum* II. Effects on population densities and grazing intensities of parrotfishes and surgeonfishes. *Marine Biology* 104:79–86.

Carrier, J. G. 1996. Marine tenure and conservation in Papua New Guinea: Problems in Interpretation. Pp. 142–167 in *The question of the Commons: The culture and ecology of communal resources*, edited by B. J. McCay and J. M. Achesons. Tucson: University of Arizona Press.

Chapin, F. S., E. Sala, I. C. Burke, J. P. Grime, D. U. Hooper, W. K. Lauenroth, A. Lombard, H. A. Mooney, A. R. Mosier, S. Naeem, S. W. Pacala, J. Roy, W. L. Steffen, and D. Tilman. 1998. Ecosystem consequences of changing biodiversity. *BioScience* 48:45–52.

Choat, J. H. 1991. The biology of herbivorous fishes on coral reefs. Pp. 120–155 in *The ecology of fishes on coral reefs*, edited by P. F. Sale. New York: Academic Press.

Choat, J. H., L. M. Axe, and D. C. Lou. 1996. Growth and longevity in fishes of the family Scaridae. *Marine Ecology Progress Series* 145:33–41.

Connell, J. H. 1978. Diversity in tropical rain forests and coral reefs. *Science* 199:1302–1310.

———. 1997. Disturbance and recovery of coral assemblages. *Coral Reefs* 16:S101–S113.

Connell, J. H., T. P. Hughes, and C. C. Wallace. 1997. A thirty-year study of coral abundance, recruitment, and disturbance at several scales in space and time. *Ecological Monographs* 67:461–488.

Cooke A., N. V. C. Polunin, and K. Moce. 2000. Comparative assessment of stakeholder management in traditional Fijian fishing-grounds. *Environmental Conservation* 27:291–299.

Cortes, J., and M. J. Risk. 1985. A reef under siltation stress: Cahuita, Costa Rica. *Bulletin of Marine Science* 36:339–356.

Cowen, R. K., K. M. M. Lwiza, S. Sponaugle, C. B. Paris, and D. B. Olson. 2000. Connectivity of marine populations: Open or closed? *Science* 287:857–859.

Cuet, P., O. Naim, G. Faure, and J. Y. Conan. 1988. Nutrient-rich groundwater impact on benthic communities of La Salina fringing reef (Reunion Island, Indian Ocean): Preliminary results. *Proceedings of the Sixth International Coral Reef Symposium* 2:207–212.

Dalzell, P. 1996. *Catch rates, selectivity and yields of reef fishing*. Pp. 161–192 in *Reef fisheries*, edited by N. V. C. Polunin and C. M. Roberts. London: Chapman and Hall.

De'ath, G., and K. E. Fabricius. 2000. Classification and regression tree: A powerful yet simple technique for ecological data analysis. *Ecology* 81:3178–3192.

Delgado, O., and B. E. Lapointe. 1994. Nutrient-limited productivity of calcareous versus fleshy macroalgae in a eutrophic, carbonate-rich tropical marine environment. *Coral Reefs* 13:151–159.

DeMartini, E. E. 1993. Modelling the potential of fishery reserves for managing pacific coral reef fishes. *Fishery Bulletin* 91:414–427.

Doherty, P. J., and D. M. Williams. 1988. The replenishment of coral reef fish populations. *Oceanography and Marine Biology* 26:487–551.

Dollar, S. J., and G. W. Tribble. 1993. Recurrent storm disturbance and recovery: A long-term study of coral communities in Hawaii. *Coral Reefs* 12:223–233.

Done, T. 1992. Constancy and change in some Great Barrier Reef coral communities: 1980–1990. *American Zoology* 32:655–662.

Done, T. J. 1987. Simulation of the effects of *Acanthaster planci* on the population structure of massive corals in the genus *Porites*: Evidence of population resilience? *Coral Reefs* 6:75–90.

———. 1988. Simulation of recovery of pre-disturbance size structure in populations of *Porites* spp. damaged by the crown of thorns starfish *Acanthaster planci*. *Marine Biology* 100:51–61.

———. 1992. Phase shifts in coral reef communities and their ecological significance. *Hydrobiologia* 247:121–132.

———. 1995. Ecological criteria for evaluating coral reefs and their implications for managers and researchers. *Coral Reefs* 14:183–192.

———. 1999. Coral community adaptability to environmental changes at scales of regions, reefs and reef zones. *American Zoologist* 39:66–79.

Done, T. J., J. C. Ogden, W. J. Wiembe, and B. R. Rosen. 1996. Biodiversity and ecosystem function of coral reefs. Pp. 393–429 in *Functional roles of biodiversity: A global perspective*, edited by H. A. Mooney, J. H. Cushman, E. Medina, O. E. Sala, and E.-D. Schulzes. New York: John Wiley and Sons.

Done, T. J., and R. E. Reichelt. 1998. Integrated coastal zone and fisheries ecosystem management: Generic goals and performance indices. *Ecological Applications* 8:S110–S118.

Drake-Brockman, J., and K. Anderson. 1994. The entwining of trade and policy in environmental issues: Implications for APEC. Honolulu: Available online at http://www.nautilus.org/papers/enviro/trade/envapec.html.

Dunne, R. P. 1994. Radiation and coral bleaching. *Nature* 368:697.

Dunne, T. 1979. Sediment yield and land use in tropical catchments. *Journal of Hydrology* 42:281–300.

Endean, R. 1976. Destruction and recovery of coral communities. Pp. 215–254 in *Biology and geology of coral reefs: Biology Two*, Vol. 3. edited by O. A. Jones and R. Endean. New York: Academic Press.

Ferreira, B. P., and G. R. Russ. 1992. Age, growth and mortality of the inshore coral trout *Plectropomus maculatus* (Pisces: Serranidae) from the Central Great Barrier Reef, Australia. *Australian Journal of Marine and Freshwater Research* 43:1301–1312.

Finn, D. 1983. Land use and abuse in the East African region. *Ambio* 12:296–301.

Furnas, M., A. Mitchell, and M. Skuza. 1997. *River inputs of nutrients and sediment to the Great Barrier Reef.* Townsville, 27–29 Nov. 1995, Great Barrier Reef Marine Park Authority (GBRMPA) Workshop Series No. 23. GBRMPA, Townsville, Queensland, Aus.

Ginsburg, N. R. 1994. *Proceedings of the Colloquium on Global Aspects of Coral Reefs: Health, Hazards and History.* Rosenstiel School of Marine and Atmospheric Science, University of Miami, Fla.

Gladfelter, W. B. 1982. White-band disease in *Acropora palmata*: Implications for the structure and growth of shallow reefs. *Bulletin of Marine Science* 32:639–645.

Glaesel, H. 1997. *Fishers, parks, and power: The socio-environmental dimensions of marine resource decline and protection on the Kenya coast.* Ph.D. diss., University of Wisconsin-Madison.

Gleason, D. F., and G. M. Wellington. 1993. Ultraviolet radiation and coral bleaching. *Nature* 365:836–838.

Glynn, P. W. 1987. Some ecological consequences of coral-crustacean guard mutualisms in the Indian and Pacific Oceans. *Symbiosis* 4:301–324.

———. 1993. Coral reef bleaching: Ecological perspectives. *Coral Reefs* 12:1–17.

———. 1997. Bioerosion and coral-reef growth: A dynamic balance. Pp. 68–113 in *Life and death of coral reefs*, edited by C. Birkeland. New York: Chapman and Hall.

Glynn, P. W., and M. W. Colgan. 1992. Sporadic disturbances in fluctuating coral reef environments: El Niño and coral reef development in the eastern Pacific. *American Zoology* 32:707–718.

Glynn, P. W., J. Cortez, H. M. Guzman, and R. H. Richmond. 1988. EI Niño (1982–83) associated coral mortality and relationships to sea surface temperature deviations in the tropical eastern Pacific. *Proceedings of the 6th International Coral Reef Symposium, Townsville* 1:237–243.

Green, A. L., C. Birkeland, R. H. Randall, B. D. Smith, and S. Wilkins. 1997. Seventy-eight years of coral reef degradation in Pago Harbor: A quantitative record. *Proceedings of the 8th International Coral Reef Symposium* 2:1883–1888.

Grigg, R. W. 1997. Paleoceanography of coral reefs in the Hawaiian-Emperor Chain—revisited. *Coral Reefs* 16:S33–S38.

Gunderson, L. H., C. S. Holling, L. Pritchard, and G. D. Peterson. Resilience in ecosystems, institutions, societies. *Beijer Discussion Paper Series No. 95.* Stockholm: Beijer International Institute of Ecological Economics.

Haas, P. M. 1997. Scientific communities and multiple paths to environmental management. Pp. 193–228 in *Saving the seas: Values, scientists, and international governance*, edited by L. A. Brooks and S. D. Van Deveer. Maryland Sea Grant program, Univ. of Maryland, College Park.

Hatcher, B. G. 1984. A maritime accident provides evidence for alternate stable states in benthic communities on coral reefs. *Coral Reefs* 3:199–204.

Hatcher, B. G., and A. W. D. Larkum. 1983. An experimental analysis of factors controlling the standing crop of the epilithic algal community on a coral reef. *Journal of Experimental Marine Biology and Ecology* 69:61–84.

Hay, M. E. 1984. Patterns of fish and urchin grazing on Caribbean coral reefs: Are previous results typical? *Ecology* 65:446–454.

Haydon, D. 1994. Pivotal assumptions determining the relationship between stability and complexity: An analytical synthesis of the stability-complexity debate. *American Naturalist* 144:14–29.

Hedley, C. 1925. The natural destruction of a coral reef. Report of the Great Barrier Reef Committee. *Transactions of the Royal Geographical Society of Australasia (Queensland)* 1:35–40.

Hodgson, G. 1997. *Resource use: Conflicts and management solutions.* New York: Chapman and Hall. Pp. 386–410.

Holland, D. S., and R. J. Brazee. 1996. Marine reserves for fisheries management. *Marine Resource Economics* 11:157–171.

Hudson, E., and G. Mace. 1996. Marine fish and the IUCN red list of threatened animals. Gland, Switzerland: IUCN (The World Conservation Union).

Hughes, T. 1993. Disturbance: Effects on coral reef dynamics. *Coral Reefs* 12:115.

Hughes, T. P. 1994. Catastrophes, phase shifts, and large-scale degradation of a Caribbean coral reef. *Science* 265:1547–1551.

Hughes, T. P., D. C. Reed, and M. J. Boyle. 1987. Herbivory on coral reefs: Community structure following mass mortalities of sea urchins. *Journal of Experimental Marine Biology and Ecology* 113:39–59.

Jablonski, D. 1998. Geographic variation in the molluscan recovery from the end-cretaceous extinction. *Science* 279:1327–1330.

Jackson, J. B. C. 1997. Reefs since Columbus. *Coral Reefs* 16:S23–S32.

Jackson, J. B. C., P. Jung, A. G. Coates, and L. S. Collins. 1993. Diversity and extinction of tropical American mollusks and emergence of the Isthmus of Panama. *Science* 260:1624–1629.

Jackson, J. B. C., A. F. Budd, and A. G. Coates. 1996. Plio-Pleistocene turnover and extinctions in the Caribbean reef-coral fauna. Pp. 169–204 in *Evolution and environment in tropical America*, edited by J. B. C. Jackson, A. F. Budd, and A. G. Coates. Chicago: University of Chicago Press.

Jennings, S., and J. M. Lock. 1996. Population and ecosystem effects of fishing. Pp. 193–218 in *Tropical Reef Fisheries*, edited by N. V. C. Polunin and C. M. Roberts. London: Chapman and Hall.

Jennings, S., and N. V. C. Polunin. 1995a. Comparative size and composition of yield from six Fijian reef fisheries. *Journal of Fisheries Biology* 46:28–46.

———. 1995b. Relationships between catch and effort in Fijian multispecies reef fisheries subject to different levels of exploitation. *Fisheries Management and Ecology* 2:89–101.

———. 1996a. Effects of fishing effort and catch rate upon the structure and biomass of Fijian reef fish communities. *Journal of Applied Ecology* 33:400–412.

———. 1996b. Impacts of fishing on tropical reef ecosystems. *Ambio* 25:44–49.

Johannes, R. E. 1978. Traditional marine conservation methods in oceania and their demise. *Annual Review of Ecology and Systematics* 9:349–364.

———. 1981. *Words of the lagoon: Fishing and marine lore in the Palau district of Micronesia.* Berkeley: University of California Press.

Kamukuru, A. T. 1992. Costs and earnings of basket trap and handline fishery in the Dar-es-Salaam region of Tanzania. Master of science thesis, Department of Applied Zoology, University of Kuopio, Finland.

Kareiva, P. M., and M. D. Bertness. 1997. Re-examining the role of positive interactions in communities. *Ecology* 78:1945.

Kauffman, S. A. 1993. *The origins of order: Self-organization and selection in evolution.* New York: Oxford University Press.

King, A. W., and S. I. Pimm. 1983. Complexity, diversity, and stability: A reconciliation of theoretical and empirical results. *American Naturalist* 122:229–239.

Kinsey, D. W. 1983. Standards of performance in coral reef primary production and carbon turnover. Pp. 209–220 in *Perspectives on coral reefs*, edited by D. J. Barnes. Manuka, Aus.: Brian Clouston Publisher.

———. 1985. Metabolism, calcification and carbon production. I. System level studies. Proceedings of the 5th International Coral Reef Congress, Tahiti. 1:505–526.

———. 1991. The coral reefs: An owner-built, high-density, fully serviced, self-sufficient housing estate in the desert—or is it? *Symbiosis* 10:1–22.

Kinsey, D. W., and P. J. Davies. 1979. Effects of elevated nitrogen and phosphorus on coral reef growth. *Limnology and Oceanography* 24:935–940.

Kinsey, D. W., and A. Domm. 1974. Effects of fertilisation on a coral reef environment—primary production studies. *Proceedings of the 2nd International Symposium* 1:49–66.

Knowlton, N. 1992. Thresholds and multiple stable states in coral reef community dynamics. *American Zoologist* 32:674–682.

Kojis, B. L., and N. J. Quinn. 1994. Biological limits to Caribbean reef recovery. A comparison with western and south Pacific reefs. Rosensteil School of Marine and Atmospheric Sciences, Miami. pp. 353–359.

Koslow, J. A., F. Hanley, and R. Wicklund. 1988. Effects of fishing on reef fish communities at Pedro Bank and Port Royal Cays, Jamaica. *Marine Ecology Progress Series* 43:201–212.

Lapointe, B. E. 1999. Simultaneous top-down and bottom-up forces control macroalgal blooms on coral reefs. *Limnology and Oceanography* 44:1586–1592.

Lapointe, B. E., M. M. Littler, and D. S. Littler. 1997. Macroalgal overgrowth of fringing coral reefs at Discovery Bay, Jamaica: Bottom-up versus top-down control. *Proceedings of the 8th International Coral Reef Symposium* 1:927–932.

Larkum, A. W. D. 1983. The primary productivity of plant communities on coral reefs. Pp. 221–230 in *Perspectives on Coral Reefs*, edited by D. J. Barnes. Manuka, Aus.: Brian Clouston Publisher.

Larkum, A. W. D., and K. Koop. 1997. Encore, algal productivity and possible paradigm shifts. *Proceedings of the 8th International Coral Reef Symposium* 1:881–884.

Larkum, A. W. D., and A. D. L. Steven. 1994. Encore: The effect of nutrient enrichment on coral reef. 1. Experimental design and research programme. *Marine Pollution Bulletin* 29:112–120.

Lessios, H. A. 1988a. Mass mortality of *Diadema antillarum* in the Caribbean: What have we learned? *Annual Review of Ecology and Systematics* 19:371–393.

———. 1988b. Population dynamics of *Diadema antillarum* (Echinodermata: Echinoidea) following mass mortality in Panama. *Marine Biology* 99:515–526.

Lessios, H. A., D. R. Robertson, and J. D. Cubit. 1984. Spread of *Diadema* mass mortality through the Caribbean. *Science* 226:335–337.

Lewis, S. A. 1986. The role of herbivorous fishes in the organization of a Caribbean reef community. *Ecological Monographs* 56:183–200.

Littler, M. M., D. S. Littler, and E. A. Titlyanov. 1991. Comparisons of N- and P- limited productivity between high granitic islands versus low carbonate atolls in the Seychelles Archipelago: A test of the relative-dominance paradigm. *Coral Reefs* 10:199–209.

Livingstone, D. A. 1996. Historical ecology. Pp. 3–17 in *East African ecosystems and their conservation*, edited by T. R. McClanahan and T. P. Young. New York: Oxford University Press.

Man, A., R. Law, and N. V. C. Polunin. 1995. Role of marine reserves in recruitment to reef fisheries: A metapopulation model. *Biological Conservation* 71:197–204.

Mapstone, B. D., R. A. Campbell, and A. D. M. Smith. 1996. *Design of experimental investigations of the effects of line and spear fishing on the Great Barrier Reef.* Technical Report No. 7. Townsville, Queensland, Aus.: CRC Reef Research Center.

Massel, S. R., and T. J. Done. 1993. Effects of cyclone waves on massive coral assemblages on the Great Barrier Reef: Meteorology, hydrodynamics and demography. *Coral Reefs* 12:153–166.

May, R. M. 1977. Thresholds and breakpoints in ecosystems with a multiplicity of stable states. *Nature* 269:471–477.

McClanahan, T. R. 1988. Seasonality in East Africa's coastal waters. *Marine Ecology Progress Series* 44:191–199.

———. 1990. Hierarchical control of coral reef ecosystems. Ph.D. diss., University of Florida, Gainesville.

———. 1992. Epibenthic gastropods of the Middle Florida Keys: The role of habitat and environmental stress on assemblage composition. *Journal of Experimental Marine Biology and Ecology* 160:169–190.

———. 1994. Kenyan coral reef lagoon fish: Effects of fishing, substrate complexity, and sea urchins. *Coral Reefs* 13:231–241.

———. 1995a. Harvesting in an uncertain world: Impact of resource competition on harvesting dynamics. *Ecological Modelling* 80:21–26.

———. 1995b. A coral reef ecosystem-fisheries model: Impacts of fishing intensity and catch selection on reef structure and processes. *Ecological Modelling* 80:1–19.

———. 1995c. Fish predators and scavengers of the sea urchin *Echinometra mathaei* in Kenyan coral-reef marine parks. *Environmental Biology of Fishes* 43:187–193.

———. 1997a. Primary succession of coral-reef algae: Differing patterns on fished versus unfished reefs. *Journal of Experimental Marine Biology and Ecology* 218:77–102.

———. 1997b. Recovery of fish populations from heavy fishing: Does time heal all? *Proceedings of the 8th International Coral Reef Symposium, Panama* 2:2033–2038.

———. 1997c. Effects of fishing and reef structure on East Africa coral reefs. *Proceedings of the 8th International Coral Reef Symposium, Panama* 2:1533–1538.

———. 1998. Predation and the distribution and abundance of tropical sea urchin populations. *Journal of Experimental Marine Biology and Ecology* 221:231–255.

———. 1999. Predation and the control of the sea urchin *Echinometra viridis* and fleshy algae in the path reefs of Glovers Reef, Belize. *Ecosystems* 2:511–523.

———. 2000. Recovery of the coral reef keystone predator, *Balistapus undulatus*, in East African marine parks. *Biological Conservation* 94:191–198.

McClanahan, T. R., K. Bergman, M. Huitric, M. McField, T. Elfwing, M. Nystrom, and I. Nordemar. 2000a. Response of fishes to algal reductions at Glovers Reef, Belize. *Marine Ecology Progress Series* 206:283–296.

McClanahan, T. R., B. A. Cokos, and E. Sala. 2002. Algal growth and species composition under experimental control of herbivory, phosphorus, and coral abundance in Glovers Reef, Belize. *Marine Pollution Bulletin* 44:441–451.

McClanahan, T. R., H. Glaesel, J. Rubens, and R. Kiambo. 1997. The effects of traditional fisheries management on fisheries yields and the coral-reef ecosystems of southern Kenya. *Environmental Conservation* 24:1–16.

McClanahan, T. R., A. T. Kamukuru, N. A. Muthiga, M. Gilagabher Yebio, and D. Obura. 1996. Effect of sea urchin reductions on algae, coral and fish populations. *Conservation Biology* 10:136–154.

McClanahan, T. R., and B. Kaunda-Arara. 1996. Fishery recovery in a coral-reef marine park and its effect on the adjacent fishery. *Conservation Biology* 10:1187–1199.

McClanahan, T. R., and S. Mangi. 2000. Spillover of exploitable fishes from a marine park and its effect on the adjacent fishery. *Ecological Applications* 10:1792–1805.

———. 2001. Comparison of closed area and beach seine exclusion on coral reef fish catches. *Fisheries Management and Ecology* 8:107–121.

McClanahan, T. R., M. McField, M. Huitric, K. Bergman, E. Sala, M. Nystrom, I. Nordemer, T. Elfwing, and N. A. Muthiga. 2001. Responses of algae, corals and fish to the reduction of fleshy frondose algae in fished and unfished patch reefs of Glovers Reef Atoll, Belize. *Coral Reefs* 19:367–379.

McClanahan, T. R., and J. C. Mutere. 1994. Coral and sea urchin assemblage structure and interrelationships in Kenyan reef lagoons. *Hydrobiologia* 286:109–124.

McClanahan, T. R., and N. A. Muthiga. 1998. An ecological shift in a remote coral atoll of Belize over twenty-five years. *Environmental Conservation* 25:122–130.

McClanahan, T. R., N. A. Muthiga, A. T. Kamukuru, H. Machano, and R. Kiambo. 1999. The effects of marine parks and fishing on the coral reefs of northern Tanzania. *Biological Conservation* 89:161–182.

McClanahan, T. R., M. Nugues, and S. Mwachireya. 1994. Fish and sea urchin herbivory and competition in Kenyan coral reef lagoons: The role of reef management. *Journal of Experimental Marine Biology and Ecology* 184:237–254.

McClanahan, T. R., and D. Obura. 1995. Status of Kenyan coral reefs. *Coastal Management* 23:57–76.

———. 1997. Sedimentation effects on shallow coral communities in Kenya. *Journal of Experimental Marine Biology and Ecology* 209:103–122.

McClanahan, T. R., and S. H. Shafir. 1990. Causes and consequences of sea urchin abundance and diversity in Kenyan coral reef lagoons. *Oecologia* 83:362–370.

McClanahan, T. R., C. R. C. Sheppard, and D. O. Obura. 2000b. *Coral reefs of the Indian Ocean: Their ecology and conservation.* New York: Oxford University Press.

McCook, L. J. 1996. Effects of water quality and herbivores on the distribution of Sargassum within fringing reefs of the central Great Barrier Reef. *Marine Ecology Progress Series* 139:179–192.

———. 1999. Macroalgae, nutrients and phase shifts on coral reefs: Scientific issues and management consequences for the Great Barrier Reef. *Coral Reefs* 18:357–367.

McGoodwin, J. R. 1990. *Crisis in the world's fisheries: People, problems, and policies.* Palo Alto, Ca: Stanford University Press.

McNaughton, S. J. 1977. Diversity and stability of ecological communities: A comment on the role of empiricism in ecology. *American Naturalist* 111:515–525.

Meyer, J. L., and E. T. Schultz. 1985. Tissue condition and growth rate of corals associated with schooling fish. *Limnology and Oceanography* 30:157–166.

Miller, M. W., and M. E. Hay. 1996. Coral-seaweed-grazer-nutrient interactions on temperate reefs. *Ecological Monographs* 66:323–344.

Moran, P. J. 1986. The *Acanthaster* phenomenon. *Oceanography and Marine Biology Annual Review* 24:379–480.

Munro, J. L. 1996. The scope of tropical reef fisheries and their management. Pp. 1–14 in *Reef fisheries*, edited by N. V. C. Polunin and C. M. Roberts. London: Chapman and Hall.

Naeem, S. 1998. Species redundancy and ecosystem reliability *Conservation Biology* 12:39–45.

Naeem, S., L. J. Thompson, S. P. Lawler, J. H. Lowton, and R. M. Woodfin. 1994. Declining biodiversity can alter the performance of ecosystems. *Nature* 368:734–737.

Sladek Nowlis, J. S., and C. M. Roberts. 1997. You can have your fish and eat it, too: Theoretical approaches to marine reserve design. *Proceedings of the 8th International Coral Reef Symposium* 2:1907–1910.

Odum, E. P. 1985. Trends expected in stressed ecosystems. *BioScience* 35:419–422.

Odum, H. T. 1983. *Systems ecology: An introduction*. New York: John Wiley and Sons.

Ohman, M. C., A. Rajasuriya, and E. Olafsson. 1997. Reef fish assemblages in northwestern Sri Lanka: Distribution patterns and influences of fishing practices. *Environmental Biology of Fish* 49:45–61.

Oliver, J., and B. L. Willis. 1987. Coral spawn slicks in the Great Barrier Reef: Preliminary observations. *Marine Biology* 94:521–529.

Ormond, R., R. Bradbury, S. Bainbridge, K. Fabricus, J. Keesing, L. DeVantier, P. Medlay, and A. Steven. 1988. Test of a model of regulation of crown-of-thorns starfish by fish predators. Pp. 190–207 in *Acanthaster and the coral reef: A theoretical perspective*, edited by R. H. Bradbury. Townsville, Queensland, Aus.: Springer-Verlag.

Pandolfi, J. M. 1996. Limited membership in Pleistocene reef coral assemblages from the Huon Peninsula, Papua New Guinea: Constancy during global changes. *Paleobiology* 22:152–176.

Patterson, K. L., D. S. Santavy, J. G. Campbell, J. W. Porter, L. G. MacGlaughlin, E. Mueller, and E. C. Peters. 1997. Coral diseases in the eastern Florida Keys, New Grounds, and the Dry Tortugas. *American Zoologist* 37:13.

Pilson, M. E. Q., and S. B. Betzer. 1973. Phosphorus flux across a coral reef. *Ecology* 54:581–584.

Pimm, S. L. 1982. *Food webs*. New York: Chapman and Hall.

Pittock, A. B., 1999. Coral reefs and environmental change: Adaptation to what? *American Zoologist* 39:10–29.

Planes, S., A. Levefre, P. Legendre, and R. Galzin. 1993. Spatio-temporal variability in fish recruitment to a coral reef (Moorea, French Polynesia). *Coral Reefs* 12:105–113.

Polovina, J. J. 1984. Model of a coral reef ecosystem: I. The ECOPATH model and its application to French frigate schals. *Coral Reefs* 3:1–11.

Polunin, N. V. C. 1984. Do traditional marine "reserves" conserve? A view of Indonesian and New Guinean evidence. *Senri Ethnological Studies* 17:266–283.

———. 1988. Efficient uptake of algal production by single resident herbivorous fish on the reef. *Journal of Experimental Marine Biology and Ecology* 123:61–76.

———. 1996. Trophodynamics of reef fisheries productivity. Pp. 361–377 in *Reef fisheries*, edited by N. V. C. Polunin and C. M. Roberts. London: Chapman and Hall.

Polunin N. V. C. and J. K. Pinnegar. 2002. Ecology of fishes in marine food webs. In *Handbook of fish and fisheries*, Vol. 1, edited by P. J. B. Hart and J. C. Reynolds. Oxford, U.K.: Blackwell.

Polunin, N. V. C., and C. M. Roberts. 1993. Greater biomass and value of target coral-reef fishes in two small Caribbean marine reserves. *Marine Ecology Progress Series* 100:167–176.

Polunin, N. V. C., C. M. Roberts, and D. Pauly. 1996. Developments in tropical reef fisheries science and management. Pp. 361–377 in *Reef Fisheries*, edited by N. V. C. Polunin and C. M. Roberts. London: Chapman and Hall.

Randall, J. E. 1967. Food habits of reef fishes of the West Indies. *Studies in Tropical Oceanography* 5:665–847.

Randall, R. H., and C. Birkeland. 1978. Guam's reefs and beaches: Part 2: Sedimentation studies at Fouha Bay and Ylig Bay. Report No. 47. Guam: University of Guam Marine Laboratory.

Rapport, D. J., and W. G. Whitford. 1999. How ecosystems respond to stress. *BioScience* 49:193–203.

Raup, D. M., and D. Jablonski. 1993. Geography of end-cretaceous marine bivalve extinctions. *Science* 260:971–973.

Reaka-Kudla, M. L., J. S. Feingold, and W. Glynn. 1996. Experimental studies of rapid bioerosion of coral reefs in the Galapagos Islands. *Coral Reefs* 15:101–109.

Risk, M. J., P. W. Sammarco, and E. N. Edinger. 1995. Bioerosion in *Acropora* across the continental shelf of the Great Barrier Reef. *Coral Reefs* 14:79–86.

Risk, M. J., and B. Sluka. 2000. The Maldives: A Nation of Atolls. Pp. 325–351 in *Coral reefs of the Indian Ocean: Their ecology and conservation*, edited by T. R. McClanahan, C. R. C. Sheppard, and D. O. Obura. New York: Oxford University Press.

Roberts, C. M. 1995. Rapid build-up of fish biomass in a Caribbean Marine Reserve. *Conservation Biology* 9:815–826.

———. 1997. Connectivity and management of Caribbean coral reefs. *Science* 278:1454–1457.

Roberts, C. M., and N. V. C. Polunin. 1991. Are marine reserves effective in management of reef fisheries? *Review in Fish Biology and Fisheries* 1:65–91.

———. 1993. Marine reserves: Simple solutions to managing complex fisheries? *Ambio* 22:363–368.

Robertson, D. R. 1991. Increases in surgeonfish populations after mass mortality of the sea urchin *Diadema antillarum* in Panama indicate food limitation. *Marine Biology* 111:437–444.

Rogers, C. S. 1993. Hurricanes and coral reefs: The intermediate disturbance hypothesis revisited. *Coral Reefs* 12:127–137.

Rose, C. S., and M. J. Risk. 1985. Increase in *Cliona delitrix* infestation of *Montastrea cavernosa* heads on an organically polluted portion of the Grand Cayman fringing reef. *Marine Ecology* 6:345–363.

Rosenau, J. N. 1997. Enlarged citizen skills and enclosed coastal seas. Pp. 329–335 in *Saving the seas: Values, scientists and international governance*, edited by L. A. Brooks and D. Van Deveer. Maryland Sea Grant Program, Univ. of Maryland, College Park.

Rowan, R., N. Knowlton, A. Baker, and J. Jara. 1997. Landscape ecology of algal symbionts creates variation in episodes of coral bleaching. *Nature* 388:265–269.

Ruddle, K. 1996. Traditional management of reef fishing. Pp. 315–335 in *Reef Fisheries*, edited by N. V. C. Polunin and C. M. Roberts. London: Chapman and Hall.

Ruddle, K., E. Hviding, and R. Johannes. 1992. Marine resources management in the context of customary tenure. *Marine Resource Economics* 7:249–273.

Russ, G. R., and A. C. Alcala. 1989. Effects of intense fishing pressure on an assemblage of coral reef fishes. *Marine Ecology Progress Series* 56:13–27.

————. 1996. Do marine reserves export adult fish biomass? Evidence from Apo Island, Central Philippines. *Marine Ecology Progress Series* 132:1–9.

Sammarco, P. W. 1980. *Diadema* and its relationship to coral spat mortality: Grazing, competition, and biological disturbance. *Journal of Experimental Marine Biology and Ecology* 45:245–272.

————. 1982. Echinoid grazing as a structuring force in coral communities: Whole reef manipulations. *Journal of Experimental Marine Biology and Ecology* 61:31–35.

Scoffin, T. P. 1993. The geological effects of hurricanes on coral reefs and the interpretation of storm deposits. *Coral Reefs* 12:203–221.

Sheppard, C. R. C. 2000. Coral Reefs of the western Indian Ocean: An overview. In *Coral reefs of the Indian Ocean: Their ecology and conservation*, edited by T. R. McClanahan, C. R. C. Sheppard, and D. O. Obura. New York: Oxford University Press.

Shulman, M. J., and D. R. Robertson. 1997. Changes in the coral reef of San Blas, Caribbean Panama: 1983 to 1990. *Coral Reefs* 15:231–236.

Simberloff, D. 1991. Keystone species and community effects of biological introductions. Pp. 1–19 in *Assessing Ecological Risks of Biotechnology*, edited by L. Ginzburg. Boston: Butterworth-Heinemann.

Smith, A., and P. Dalzell. 1993. Fisheries resources and management investigations in Woleai Atoll, Yap State, Federated States of Micronesia. (4). Inshore Fisheries Research Project Technical Document, pp. 64.

Smith, A. H., and F. Berkes. 1991. Solutions to the "Tragedy of the Commons": Sea-urchin management in St. Lucia, West Indies. *Environmental Conservation* 18:131–136.

Smith, S. V. 1983. Coral reef calcification. Pp. 240–247 in *Perspectives on coral reefs*, edited by D. J. Barnes. Manuka, Aus.: Brian Clouster Publisher.

Smith, S. V., W. J. Kimmerer, E. A. Laws, R. E. Brock, and T. W. Walsh. 1981. Kaneohe Bay sewage diversion experiment: Perspectives on ecosystem responses to nutritional perturbation. *Pacific Science* 35:279–402.

Smith, T. D. 1988. Stock assessment methods: The first fifty years. Pp. 1–34 in *Fish population dynamics*, edited by J. A. Gulland. Chichester: John Wiley and Sons.

Spear, T. T. 1978. The kaya complex: A history of the Mijikenda peoples of the Kenya coast to 1900. Nairobi: Kenya Literature Bureau.

Stanley, S. M. 1986. Anatomy of a regional mass extinction: Plio-Pleistocene decimation of the Western Atlantic bivalve fauna. *Palaios* 1:17–36.

Steneck, R. S., and M. N. Dethier. 1994. A functional group approach to the structure of algal-dominated communities. *Oikos* 69:476–498.

Sweatman, H. 1997. *Long-term monitoring of the Great Barrier Reef*. Status Report No. 2. Australian Institute of Marine Science, Townsville, Queensland, Aus.

Szmant, A. M. 1997. Nutrient effects on coral reefs: A hypothesis on the importance of topographic and trophic complexity to reef nutrient dynamics. *Proceedings of 8th International Coral Reef Symposium* 2:1527–1532.

Tanner, J. E. 1995. Competition between scleractinian corals and macroalgae: An experimental investigation of coral growth, survival and reproduction. *Journal of Experimental Marine Biology and Ecology* 190:151–168.

Thomson, D. B. 1980. Conflict within the fishing industry. *ICLARM Newsletter* 3:3–4.

Thompson, M., and A. Trisoglio. 1997. Managing the unmanageable. Pp. 107–127 in *Saving the sea: Values, scientists and international governance,* edited by L. A. Brooks and D. Van Deveer. Maryland Sea Grant program, Univ. of Maryland, College Park.

Tilman, D., and J. A. Downing. 1994. Biodiversity and stability in grasslands. *Nature* 367:363–365.

Tilman, D., D. Wedin, and J. Knops. 1996. Productivity and sustainability influenced by biodiversity in grassland ecosystems. *Nature* 379:718–720.

Tomascik, T., A. J. Mah, A. Nontji, and M. K. Moosa. 1997. *The ecology of the Indonesian seas.* Hong Kong: Periplus.

Tomascik, T., and F. Sanders. 1987. Effects of eutrophication on reef building corals II. Structure of scleractinian coral communities on fringing reefs, Barbados, West Indies. *Marine Biology* 94:63–75.

Tomascik, T., R. v. Woesik, and A. J. Mar. 1996. Rapid coral colonization of a recent lava flow following a volcanic eruption, Banda Islands, Indonesia. *Coral Reefs* 15:69–176.

Tribble, G. W., M. J. Atkinson, F. J. Sansone, and S. V. Smith. 1994. Reef metabolism and end-upwelling in perspective. *Coral Reefs* 13:199–201.

Umar, J., L. J. McCook, and I. R. Price. 1998. Effects of sediment deposition on the seaweed *Sargassum* on a fringing coral reef. *Coral Reefs* 17:169–177.

UNEP/IUCN (United Nations Environmental Programme/The World Conservation Union). 1988. *Coral reefs of the world.* Gland, Switzerland: IUCN/UNEP.

van Katwijk, M. M., N. F. Meier, R. Loon, E. M. Hove, W. B. J. T. Giesen, G. Velde, and C. Hartog. 1993. Sabaki river sediment load and coral stress: Correlation between sediments and condition of the Malindi-Watamu reefs in Kenya (Indian Ocean). *Marine Biology* 117:675–683.

Vermeij, G. J. 1978. *Biogeography and adaptation: Patterns of marine life.* Cambridge, Mass.: Harvard University Press.

Veron, J. E. N. 1995. *Corals in space and time: The biogeography and evolution of the scleractinia.* Sydney, Aus.: UNSW Press.

Wilkinson, C. R., D. M. Willams, P. W. Sammarco, R. W. Hogg, and L. A. Trott. 1984. Rates of nitrogen fixation on coral reefs across the continental shelf of the central Great Barrier Reef. *Marine Biology* 80:255–262.

Williams I. D., N. V. C. Polunin. 2000. Differences between protected and unprotected Caribbean reefs in attributes preferred by dive tourists. *Environmental Conservation* 27:382–391.

———. 2001. Large-scale associations between macroalgal cover and grazer biomass on mid-depth reefs in the Caribbean. *Coral Reefs* 19:358–366.

Williams I. D., N. V. C. Polunin, and V. Hendrick. 2001. Limits to grazing by herbivorous fishes and the impact of low coral cover on macroalgal abundance on a coral reef in Belize. *Marine Ecology Progress Series* 222:187–196.

Williams, S. L., and R. C. Carpenter. 1988. Nitrogen-limited primary productivity of coral reef algal turfs: Potential contribution of ammonium excreted by *Diadema antillarum. Marine Ecology Progress Series* 47:145–152.

Woodley, D. J., E. A. Chornesky, P. A. Clifford, J. B. C. Jackson, L. S. Kaufman, N. Knowlton, J. C. Land, M. P. Pearson, J. W. Porter, M. C. Rooney, K. W. Rylaarsdam, V. J. Tunnicliffe, C. M. Wahle, J. L. Wulff, A. S. G. Curtis, M. D. Dallmeyer, B. P. Jupp, M. A. R. Koehl, J. Niegel, and E. M. Sides. 1981. Hurricane Allen's impact on Jamaican coral reefs. *Science* 214:749–755.

Young, T. P. 1996. High montane forest and afroalpine ecosystems. Pp. 401–424 in *East African ecosystems and their conservation*, edited by T. R. McClanahan and T. P. Young. New York: Oxford University Press.

Zarsky, L., and J. Hunter. 1998. Communities, markets and city government: Innovative roles for coastal cities to reduce marine pollution in the Asia-Pacific. Berkeley, Calif.: Nautilus Institute. Available online at http://www.nautilus.org/papers/enviro/hunterzarsky_marine.txt.

Zerner, C. 1994. Tracking sasi: The transformation of a central Moluccan reef management institution in Indonesia. Pp. 19–32 in *Collaborative and community-based management of coral reefs: Lessons from experience*, edited by A. T. White, L. Z. Hale, Y. Renard, and L. Cortes. West Hartford, Conn.: Kumarian Press.

6

Resilience in Wet Landscapes of Southern Florida

Lance H. Gunderson and Carl J. Walters

The wet ecosystems of southern Florida provide good examples of ecological resilience. Understanding resilience in those ecosystems has both theoretical and practical implications. During this century in southern Florida, human population growth has been phenomenal, increasing from less than ten thousand to more than 4 million. That growth has brought large-scale changes in land use patterns and concomitant changes in ecosystem structure and function. Among the variety of human uses imposed on the southern Florida peninsula is the desire to conserve "natural" ecosystems—either in preserves and conservation areas or as national parks. To that end, about one-half of the historic Everglades ecosystem has been set aside for some form of conservation (Gunderson and Loftus 1993). That area of remnant Everglades is the subject of attempts to restore lost values around the notion of ecosystem restoration (Davis and Ogden 1992).

The remainder of this chapter is divided into four sections. In the first section, a heuristic model is presented to provide a framework for discussing the property of resilience. This framework is used to explain resilience in two systems of southern Florida—the freshwater marshes of the Everglades as described in second section and in Florida Bay in the third. Florida Bay is the shallow marine lagoon at the tip of the peninsula (figure 6.1). The last section discusses some management implications through the optic of ecosystem resilience.

Another Heuristic of Resilience

The introductory chapters of this volume provide a theoretical background and definitions for resilience. Those chapters enrich the discussion by using

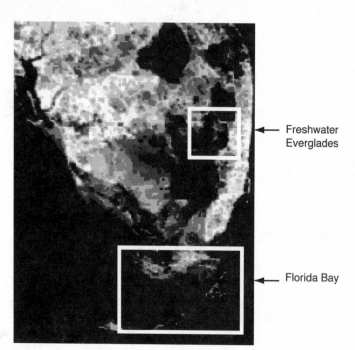

Freshwater Everglades

Florida Bay

Figure 6.1. Map indicating location of two South Florida ecosystems that have lost ecological resilience. The freshwater Everglades region that was once dominated by sawgrass (*Cladium jamaicense*) is now dominated by cattail (*Typha* sp.), indicating a shift in stability regime. Florida Bay was dominated by clear water and seagrass until 1991, when a massive die-off occurred, followed by a transition to turbid water and a phytoplankton-dominated system.

metaphors to describe resilience, including verbal descriptions and mathematical representations (Ludwig et al. chapter 2). To reiterate, resilience is contrasted with stability (*sensu* Holling 1973), in that stability is the tendency of a system to return to an equilibrium following a disturbance, while resilience is defined as the amount of disturbance that a system can absorb before it changes stability domains.

To further highlight differences between definitions of resilience and stability we'll use a heuristic model—that of a ball in a cup. Picture a ball or marble sitting in the bottom of a cup on a table (figure 6.2a). The ball is stable (it doesn't move) and its resting point is at an equilibrium created by a balance of the downward force of gravity and the upward force of the table and cup. If the cup is shaken, the ball will soon return to an equilibrium position. This physical model captures the essence of engineering resilience; a global equilibrium with resilience defined as the time taken for the ball to return to a stationary position.

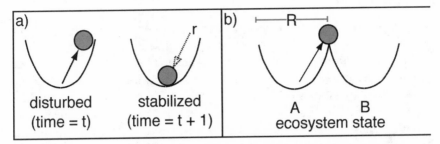

Figure 6.2. Alternative definitions of resilience as represented by ball-and-cup diagrams. *(a)* Engineering resilience sensu Holling (1995) is the time of return following a disturbance to the system, and *(b)* ecological resilience is the amount of disturbance that a system can absorb without changing stability regimes or alternative ecosystem states.

To understand ecological resilience we must change one ingredient of the model: the cup. In ecological resilience, the cup is not rigid, nor is there just one cup; there are multiple cups on the table (figure 6.2b), and the cups morph or change shape over time. In a landscape of multiple, morphing cups, a ball can easily move from one cup to another—change stability domains. Also, the changing landscape allows for a smaller disturbance to cause the ball to move among cups. This is analogous to the loss of resilience in ecological systems. That is, the system is no longer capable of absorbing a disturbance, and so the disturbance suddenly flips the system into another stability domain. This property is referred to as *adaptive capacity* (Gunderson et al. chapter 1).

Each of the pieces in the cup metaphor represents key variables that operate at characteristically different speeds. The ball can move quickly and represents a rapidly changing variable. The shaking or disturbance occurs at a different time interval and represents variable(s) that operate at slower speeds (measured by longer-term variations, longer cycle times). The stability domain is defined by the characteristics of the cup—the slowest-changing variable in the model. In ecologic systems the stability regime is defined by slowly changing variables. Hence it is the dynamics of the slowly changing variables that interact with faster variables (disturbances, and other key structural elements) which define ecological resilience. We build on this cross-scale interaction of variables in the next section to explain how resilience and stability change over time, resulting in qualitative changes in ecosystem behavior and properties. We look at two examples from wet landscapes in southern Florida, starting with the freshwater Everglades.

Freshwater Everglades

One example of the loss of landscape resilience involves nutrient enrichment in the freshwater marshes of the Everglades. The Everglades is an oligotrophic wetland, limited primarily by phosphorus (Steward and Ornes 1973). For the past five thousand years or so, the ecosystem effectively self-organized around this low nutrient status, pulsed by annual wet/dry cycles and by decadal recycling associated with fires (Loveless 1959; Craighead 1971; Gunderson 1994). The resulting landscape mosaic had small areas of enhanced nutrients or eutrophy in tree islands that were maintained by wading bird nesting or in local refugia maintained by the cycles of flooding and drydowns that first collected diffuse energy and then concentrated it locally. The remainder of the landscape (sawgrass marshes and wet prairies) adapted to low nutrient thresholds (Steward and Ornes 1973).

In the late 1940s a plan was put into effect that divided the Everglades into three designated land uses: agriculture (in the northern third of the historic Everglades), urban (the eastern fifth), and conservation (in the southern and central remaining half of the historic system). The latent effects of these land use designations were revealed in the late 1970s and early 1980s, when large-scale shifts in the vegetation were noticed in the areas immediately south of the agricultural area. After years of research, the transition from a sawgrass- to cattail-dominated marsh was attributed to a two-part process. First, there was a slow increase in the concentration of soil phosphorus levels (Davis 1989, 1994) which was followed by a fire, drought, or freeze, resulting in the replacement of sawgrass by cattail (Newman et al. 1997, 1998).

Each of these components represents processes and structures that occur at different spatial and temporal scales. The vegetation structures represent the most rapidly changing variables, with plant turnover times on the order of five to ten years (Davis 1989, 1994). The disturbance regimes of fires operate on return frequencies of ten to twenty years (Wade et al. 1980; Gunderson 1994). Other disturbances such as freezes and droughts occur on multiple-decade return times (Gunderson 1992; Duever et al. 1994). The soil phosphorus concentrations are the slowest of the variables, with turnover times on the order of centuries (Davis 1994).

The resilience of the freshwater marshes is related to the soil nutrient content. That is, it is the soil phosphorus level that determines the stability landscape or the adaptive capacity of the system to deal with disturbances. These disturbances are in the form of external variations in rainfall, temperature, and fires. The alternative stability domains are characterized by the dominant plant species: sawgrass and wet prairie communities dominate on sites with low nutrients, and cattail dominates on sites with higher soil phosphorus concentrations.

It is the slowly changing soil phosphorus level that determines the ability of the system to absorb disturbances.

Fires, droughts, and freezes have all been part of the configuring processes around which the Everglades has self-organized for over fifty centuries. Charcoal layers are present throughout the peat profiles (Cohen 1984), evidence of the occurrence of fire throughout the recent history of the Everglades. The original stability domain of sawgrass and wet prairie matrix was maintained by the interaction of fire, droughts, and subsequent hydrologic regimes (Herndon et al. 1991). When severe, peat-removing fires would lower soil elevations, then a wet prairie community would replace previous sawgrass communities (Craighead 1971). Without recurrent, severe fires, the sawgrass community would slowly colonize the wet prairie community. The composition of this shifting matrix had little or no cattails, except around alligator holes or other areas where animals concentrated nutrients. It is only since the late 1970s and early 1980s that cattail has formed large monotypic stands in areas of nutrient enrichment (Davis 1994).

To summarize to this point, the vegetation shifts in nutrient-enriched areas of the freshwater Everglades can be explained by the interactions of three sets of variables. The most rapidly changing variable set is the dominant plant species, characterized by either sawgrass in the low nutrient state or cattail in the high nutrient state. Any of a number of disturbances (fires, droughts, freezes) can trigger a switch between these states—a phenomenon that has been the subject of much scientific debate and legal maneuvering. The third key variable is the soil phosphorus content, which defines the alternative stability domains. A similar typology can be used to explain the dynamics of a shift in stability domains in Florida Bay.

Florida Bay

Florida Bay is a shallow, subtropical marine ecosystem located at the southern terminus of the Florida peninsula. The bay is ecologically linked to the north with the freshwater Everglades, to the west with the Gulf of Mexico, and to the south and east with the Florida Keys and Atlantic Ocean. Since the late 1980s, the bay ecosystem has undergone a series of dramatic changes including massive die-off of seagrass (Robblee et al. 1991), recurring blooms of phytoplankton, and altered aquatic communities. Although change is part of all ecological systems, these changes have altered the essential character of the bay; the system has flipped from being a clear-water, grass-dominated system to one with less seagrass, turbid water, and recurring algae blooms. Each of these states is stable over time.

Florida Bay is subject to recurrent disturbances such as tropical storms, hurricanes, freezes, and periodic pulsing of freshwater from the wetlands of the southern peninsula. These climatic conditions result in variable salinity and temperature patterns in the bay. The bay system is also oligotrophic, primarily phosphorus limited (Fourqurean et al. 1993). Understanding the role of these factors is two-fold: (1) in determination of what conditions created the flip in stability domains, and (2) in development of considered management actions, with a goal of restoring the ecosystem from a turbid, yellow-green to the socially and economically desirable clearer water and seagrass.

Many explanations for the observed flip in the state of the bay have been proposed. The proposed explanations are arguments of a local surprise—where variation at the local system level (driven by external processes) has exceeded previously observed bounds. The explanations include an increase in the variability of salinity, an increase in nutrient input, and a decrease in recurrent severe storms. The alternative or competing explanations are described in the following paragraphs, as caricatures of the Bay.

The Senile Bay.	No major hurricanes have impacted the Bay since 1965. Hurricanes destroy seagrass beds, move sediments around, and "flush" the bay. Without hurricanes, the seagrass beds become overgrown and vulnerable to die-off.
The Thirsty Bay.	Changes in freshwater flow from the Everglades have occurred this century due to upstream water management. Lowering the volume of fresh water entering the bay and diverting entry points have allowed salinity to increase in the bay, which in turn have contributed to the die-off of seagrass and inhibited reestablishment.
The "Outhouse" Bay.	Nutrients have increased in the Bay due to land use changes such as development in the Keys and elsewhere in Florida. Seagrass thrives under conditions of low nutrients, while certain types of algae utilize higher nutrient concentrations.
The "Topless" Bay.	Changes in key animals populations have also occurred in this century. Large grazers such as sea turtles and manatees would maintain lower seagrass biomass with more spatial heterogeneity, making it less vulnerable to die-off processes.
The Stagnant Bay.	Construction activities of the overseas railroad filled in passes among some of the islands of the Florida Keys. As early as 1920, these closures restricted circulation in the

The Rising Bay. bay, resulting in lower turnover in the northeastern portions.

The Florida peninsula is relatively stable geologically; however, the sea level has risen on the order of 30 centimeters this century. If not offset by marl accretion, the bay may be deeper, with resulting changes in light penetration and circulation.

These hypotheses are not exclusive; the actual changes may have been caused by a combination of these mechanisms. The alternative hypotheses are important because subsequent management actions are dramatically and diametrically different depending upon which of the above hypotheses might be true. For example, if nutrient additions have resulted in the observed changes, then adding more freshwater may not change the stability regime of the bay.

To begin sorting among the competing hypotheses, we developed a set of frameworks for iterating among observations, data, and explicit hypotheses (models). This approach not only integrates and synthesizes understanding, but also allows for clearer articulation of certainties and uncertainties. We organized the inquiry in two frameworks: one to examine circulation, freshwater, and salinity interactions in the bay and the other to examine seagrass dynamics.

Circulation, Freshwater, and Salinity

Broad-scale and long-term (nontidal) circulation over the Bay is driven by external sea surface anomalies coupled with variable wind speed and direction. The shallow, complicated bathymetry of the bay interacts with these drivers to produce four distinct areas of circulation. The western bay has the highest currents tending to move from northwest Gulf of Mexico to southeast through passes between the middle Florida Keys. The north-central bay has restricted circulation due to a series of mud embankments but has more interchange with the west than does northeast Florida Bay, where little long-term flow exists. Water in the south-central portion of the Bay (Lignumvitae Basin) connects to and circulates with the Atlantic Ocean. These areas also have been described has having distinct faunal assemblages (Ginsberg and Lowenstam 1958).

Starting in the 1920s, a number of changes have occurred in the upstream hydrology that have decreased the amount of overland freshwater entering the bay. These include canal and levee construction, and changes in operational rules. Two major flowways influence the bay: Shark Slough and Taylor Slough. Shark Slough is the larger of the two (historic flows on the order of 2-million-acre feet per year) and enters the Gulf of Mexico to the north and west of the bay. Taylor Slough contributes less (roughly one-quarter of Shark Slough volumes) and delivers water primarily to northeast Florida Bay.

At times, Shark Slough has a dramatic effect on changing the amplitude of salinity fluctuations in the western bay. The prevailing currents move freshwater relatively rapidly from Shark Slough around Cape Sable and into the western part of the bay. Yet even with this influence of freshwater flow, consistent, recurrent patterns of hypersalinity occur in the west-central bay (Garfield-Rankin area). Another, longer-term salinity reconstruction using coral growth in the bay by Swart et al. (1996) indicates similar patterns to those predicted by our models: no long-term trend and a complicated variation over time—linked to the varying degree of interaction between upstream flow, and in situ rainfall/evaporation.

These reconstructions suggest that periods of extreme hypersalinity have happened in the past and are likely to happen again. If correct, and the bay has a recent history of a highly variable (in space and time) but trendless salinity regime, why have large-scale seagrass die-off and flips in ecosystem structure not been recorded before? This question is addressed in the next section, where linkages between seagrass stand dynamics and environmental variables are discussed.

Seagrass Community Dynamics

In order to sort among the competing explanations regarding seagrass community die-off and subsequent shifts, we qualitatively examine changes in disturbance regimes that may have affected seagrass structure, followed by development of a model to examine factors that may have led to the observed state changes.

Seagrass and algae-free water have characterized the bay since the early 1960s (Tabb et al. 1962; Ginsberg and Lowenstam 1958), a period relatively free of broad-scale disturbances. Prior to that time, hurricanes, grazers, and humans all acted to recurrently decrease seagrass biomass. During this century, hurricanes entered Florida Bay in 1906, 1926, 1935, 1947, 1960 (Donna), and 1965 (Betsy). Hurricanes Donna and Betsy restructured sediment distributions, (due to supratidal inundation—the recorded range during Donna was 13-plus feet, normal is less than 1 foot), and both storms decreased the standing seagrass biomass (Perkins and Enos 1968; Ball et al. 1967). Hurricane Andrew passed north of the Bay in 1992 with no local effect, but it did however create a pulse of nutrient input associated with widescale upstream defoliation of mangroves and subsequent run-off (Tilmant et al. 1994). Populations of sea turtles and manatees have probably not been sufficient since the 1960s to create large-scale grazing impacts on seagrass standing stock. In the early 1980s, commercial netting was removed from the park portions of the bay, as were trawling fishermen during the 1950s. The net result is that since the 1950s the types of disturbances that would decrease seagrass biomass have all lessened.

However, it remains an open question as to whether the multi-decade lack of disturbances contributed to the die-off and subsequent shifts. A set of arguments for the seagrass die-off period focus on external variations to the system—or on proximal triggers: metabolic stress due to hypersalinity, temperature, and decreased light levels (Robblee et al. 1991). Durako and Kuss (1994) report on the role of *Labyrinthia* as a disease agent. Carlson (1994) indicated that hydrogen sulfide is a secondary, synergistic stressor to seagrass. Since the die-off primarily affected *Thalassia* beds in areas of relatively high biomass (Robblee et al. 1991), we turned our focus to a framework to evaluate the role of standing stock as a key factor in defining the resilience of the system.

Modeling Seagrass Dynamics

We developed a model to examine the possibility that development of seagrass biomass in some areas of Florida Bay is directly limited by oxygen supply for respiration and that buildup of high biomass may increase vulnerability to an anoxic stress syndrome leading to die-offs at basin scales. A number of observations lead us to suspect such a syndrome. The most lush beds are found in areas of strong current and hence high oxygen delivery rates to the bottom. Die-offs were associated with apparent oxygen depletion as evidenced by occurrence of fish die-offs in areas like Johnson Key Basin. In lush beds, simple calculations of daily oxygen production, respiration, and diffusion from the atmosphere indicated a potential for strong imbalance (respiration > production + diffusion) should any factor cause even modest decrease in primary production rates. Die-offs were associated with periods of high temperature and salinity, both of which substantially decrease oxygen saturation concentrations and hence potential delivery rates of oxygen from atmospheric sources to the bottom.

The model consists of two parts: (1) an oxygen balance model accounting for rates of primary production, respiration, atmospheric exchange of O_2 with the water column, and diffusion of O_2 from the water column into the neighborhood of the bottom leaf zone; and (2) a biomass/detritus balance model accounting for changes in live biomass and detritus with changes in primary production and respiration rates, and including a strong effect of oxygen concentration at the leaf zone on respiration rates.

Primary production, respiration, and leaf loss rates (root loss rates assumed comparable) per unit biomass for the analysis were estimated from data in Zieman et al. (1989), Durako and Kuss (1994), and Bosence (1989). For total biomass rather than leaf biomass only assessment, the leaf/root biomass ratio was taken to be around 5:1, rates per leaf biomass per day were assumed to be on the order of 0.05 grams per gram per day primary production, 0.02 grams per gram per day leaf/root loss, and 0.03 grams per gram per day respiration (for a balanced bed with no net production going to biomass increase). Decomposition rates of detritus were assumed to average 0.03–0.06 grams per gram per day. The exact values of these rates were not critical for the analysis. Primary production rate was assumed nearly proportional to biomass as suggested by Zieman et al. (1989), and leaf loss rate was assumed to be inversely proportional to oxygen concentration such that the modeled loss rate increased to 0.04 grams per gram per day if oxygen concentration in the water column dropped to 3 parts per million (and 0.08 grams per gram per day at 1.5 parts per million, etc.).

We input monthly time series reconstructions of temperature and salinity for the period 1960–1995, where the monthly values of these driving variables influenced oxygen saturation and reaeration rate. Monthly data on variation in wind speed indicated that it would not be worthwhile to also include wind speed effects on reaeration rates. For some tests, we also assumed primary production rates and respiration rates to be proportional to temperature, to generate seasonal changes in both supply and biological demand for oxygen.

The model predicts a wide variety of dynamic behaviors over time, with violent changes in behavior due to small changes in parameter values representing effects such as water circulation rate and depth-light penetration effects on primary production (figure 6.3). For most parameter combinations (sites in Florida Bay), the model predicts either stable or persistently erratic biomass changes. For a few combinations, it predicts regular, cyclic biomass development and collapse. And for a few other parameter combinations, it predicts long-term biomass development and rare catastrophic die-offs with the die-offs generally timed to coincide with historical periods of high temperature and salinity that have followed multiyear periods favorable for biomass accumulation.

Thus the model can "explain" die-offs in particular areas of Florida Bay without invoking external stress agents such as disease or reduced primary production due to phytoplankton blooms, and without invoking possibly unrealistic assumptions about the sensitivity of *Thalassia* to salinities in the range 20–50 parts per thousand. In scenarios where the model predicts catastrophic die-offs at about the time when these occurred in the field, the basic mechanism involves a "runaway" hypoxic stress syndrome: after a period of several years favorable for biomass development (moderate salinities permitting good aerobic respiration

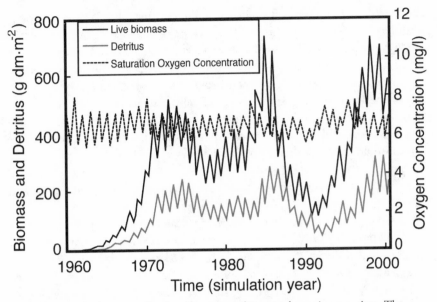

Figure 6.3. Simulation results showing seagrass biomass dynamics over time. The interaction among the oxygen production from live biomass, the saturation concentration of oxygen in the water column, and the respiratory demand of standing stock (live and dead biomass) can generate a rapid decline in live biomass on a multi-decadal time span.

rates and little loss of biomass to anaerobic respiration), a high salinity/temperature situation is predicted to cause decreases in potential oxygen delivery rate to the bottom community. This causes a decrease in oxygen concentration, which triggers inefficient anaerobic respiration and increased leaf/root mortality rates. The increased mortality rates cause primary production (large component of oxygen available at the bottom) to decrease and detritus to increase, leading to a strong temporary imbalance between oxygen production and demand. Oxygen concentration in the water then decreases rapidly, further increasing mortality rates and reducing primary production. Oxygen concentrations recover very quickly (within a few months) after such spirals of biomass decline and detritus production, but biomass takes much longer to recover.

Iteration between models and data has helped to clarify arguments surrounding the competing hypotheses. A common thread through all of these hypotheses is seagrass biomass. That is, all of the hypotheses focus on the spatial and temporal dynamics of biomass and, more importantly, on those aspects of biomass that generate a vulnerability to the kind of anoxic stress that creates a runaway die-off. The die-off event was unique in that there was no previous record of such a large-scale mortality (Robblee et al. 1991). This suggests that the key variable controlling such a process had similar attributes over a similarly broad scale, or that the system was spatially homogeneous. A plausible explanation that could lead to such a homogeneous system is the removal of, or lessening of the frequency of, disturbances. These disturbances, such as large-scale hurricanes and small-scale agents (mullet, sea turtles) would probably have differential effects and result in a spatial mosaic. Such a mosaic of different mass stands would be less likely to "carry" a die-off.

Since the massive die-offs of the late 1980s, a number of exogenous events have influenced seagrass recolonization of the Florida Bay. Following the initial die-off, Thayer et al. (1994) reported biomass in many seagrass beds had increased by 1992—indicating recovery. Hurricane Andrew passed north of the bay in August 1992, defoliating mangrove forests along the bay's northern boundary. That biomass was subsequently flushed into the bay, adding to nutrient loads. Even with that nutrient pulse, the seagrass biomass was still increasing in 1993 (Durako pers. comm.). In 1993, severe flooding occurred in the Mississippi River basin, resulting in a large pulse of nutrients into the Gulf of Mexico. Indicator contaminants associated with the floodwater were noticed within a few weeks in the Florida Keys, suggesting that possibility of additional nutrient inputs from Gulf waters into the bay. The point here is that distant events, over which managers have no control, have influenced and will likely continue to influence proximal management actions.

Models, data, and field observations all indicate that the seagrass beds are recovering, although algae blooms and turbidity plumes still occur and perhaps continue to impede recovery. That recovery is characterized by an approximate decade-long time lag since the first observed die-off events. The trajectory of recovery appears headed toward a clear water system, although chronic turbidity from marl resuspension and plankton may hinder return to a completely clear state. Shark Slough discharges also seem to be a source of recurrent sediment plumes that move around Cape Sable and into the bay (National Oceanic and Atmospheric Administration satellite data), creating turbid conditions that limit recovery. In other words, well meaning yet singular quick fixes such as delivering more freshwater into the bay may be impeding efforts rather than facilitating seagrass recovery.

To summarize the Florida Bay story, the alternative stability domains are characterized by clear water/seagrass beds and turbid water with phytoplankton. The resilience of the bay is characterized by the interaction of key variables, each operating at characteristic time scales. Seagrass biomass is the slowly changing (decadal scale) variable at the center of the alternative hypotheses.

Table 6.1. Key variables, rates, and stability domains of ecosystems in South Florida

System	Key Variable	Rates	Stable State 1	Stable State 2
Freshwater Everglades Marsh	Species Composition	Fast	Sawgrass	Cattail
	Disturbances: Fire, Frost, Drought	Intermediate		
	Soil Phosphorus	Slow		
Florida Bay	Primary Producers	Fast	Seagrass	Phytoplankton
	Disturbances	Intermediate		
	Connectivity	Slow		
	Seagrass			
	Biomass			

The oxygen requirements of that biomass and the suite of mechanisms that provide oxygen are determinants of vulnerability. Hence, well-aerated areas (channel bottoms, basins with better circulation) are less vulnerable to disturbances that would generate the autocatalytic die-offs observed there in the late 1980s. The role of salinity was surprising, not as a stress agent but as a determinant of oxygen saturation in the water column. The ecosystem behavior, characterized by cusps and hysteresis, has been demonstrated in semi-arid grazing systems (Walker et al. 1981), in homoclinic bifurcation dynamics observed in shallow lakes in the Netherlands (Scheffer et al. 1993), and in primitive models of resilience (Ludwig et al. 1997).

The key variables in both of these examples operate at distinctly different scales and can be mapped into the heuristic of resilience (table 6.1). In both cases, the plant species composition is the "fast" variable, which has shifted stability domains. In the Everglades, the dominant taxa has changed from sawgrass to cattail; in the bay, it is a shift from the benthic seagrass to the phytoplankton in the water column. In both of these examples, a disturbance is key in causing a flip; in the Everglades, fires, droughts, or freezes can be the trigger, whereas in the bay it appears that high temperatures and salinities perturbed the system. Finally, it is the changes in "slow" variables that define the loss of resilience. For the Everglades, it was the slow, gradual accretion of soil phosphorus, and for the bay it was the accumulation of standing biomass in the seagrass beds. In each of these events, understanding the property of resilience was key to robust management actions.

Management Implications

Ecosystem state changes generate surprises for scientists and great uncertainties for stakeholders and policy makers. Generally, scientists or technical experts can develop a number of plausible explanations or competing hypotheses for the observed ecosystem changes. Three sets of actions seem to be most commonly adopted in response to the uncertainties of the ecosystem dynamics. The first is to implement a set of singular management actions to fix the apparent problem. The second is to adopt one of the hypotheses, because other motives (social or political) may be affiliated with the policy implications of a preferred hypothesis. The third is to develop an integrated assessment that can independently sort through the alternative explanations to develop plausible management probes to test the uncertainties.

Myopic management actions that are designed to "fix" a perceived problem usually backfire or create unintended responses. One example is from Florida Bay. A group concluded that the observed state changes in the bay were a result of increased salinity due to decreased overland freshwater flow. A policy was enacted to increase that flow into and through Shark Slough in order to ameliorate hypersalinity in the bay. The policy performed as intended and more water was delivered into the headwaters just northwest of the bay. However, while salinities were moderated, the increased flow generated turbidity plumes that moved in the bay, decreasing light penetration and inhibiting seagrass recolonization (Walters 1997). These fixes that backfire usually focus on a single variable, a model of simple cause and effect, and an implicit view of single, global equilibrium.

A more insidious response is where stakeholders and vested interests that benefit from existing policies use the uncertainties generated by competing hypotheses to maintain the status quo. That is, these interests use the alternative hypotheses to create and maintain an uncertainty of action. This was the case in the nutrient enrichment of the freshwater Everglades. Scientists funded by farmers initially proposed arguments of changes in water depths, flow, and disturbances to explain the vegetation shifts from sawgrass to cattail. A set of similar responses is chronicled in Ehrlich and Ehrlich (1993) around issues of loss of biodiversity, global warming, and carcinogenic effects of tobacco. Independent science, removed as far from politics as possible, is a key ingredient in breaking through this type of obfuscation and stalemate.

An integrated assessment is crucial to resolution of competing hypotheses regarding a resource issue—especially where the observed ecosystem state changes create the surprise. Articulation of alternative or competing explanations helps to make the uncertainties explicit. An integrated assessment provides the mechanism for sorting among these competing explanations. The workshops and models developed for the adaptive environmental assessment and management process are used to iterate among data, observations, and models in order to reject unfeasible hypotheses (Walters 1986).

The manner of resolution of the competing hypotheses dictates subsequent actions. In the best (and most rare) of all situations, all but one of the competing hypotheses can be rejected, leaving only one hypothesis that can be put at risk by a set of management actions. But in many cases such winnowing is not possible, yet policy probes can still be structured because the competing hypotheses can be tested with similar types of management actions. Such was the case in the competing hypotheses regarding decline in wading bird nesting in the Everglades—where at least the three of the hypotheses (disruption in timing of feeding, disruption in food delivery, loss of estuarine productivity) could all be tested by hydrologic manipulation (Walters et al. 1992). Yet even these actions were constrained because of stakeholder inflexibility.

A central tenet of the adaptive management approach to dealing with competing hypothesis is to structure management probes for learning. Yet learning seems to be intertwined with cycles of policy success and failure (Westley 1995). If policies are working (or appear to be working), there is little or no emphasis on learning (why should one learn from success?). It is when policy fails, either dramatically or chronically, that learning is deemed a priority and necessary. The challenge to develop a capacity for learning continues to be problematic among many western resource management institutions. Yet when needed, that capacity seems to be come from focusing on understanding (not efficiency) and by networking with those who practice learning.

Concluding Remarks

Perhaps it is time to rethink the paradigms or foundations of resource management institutions and to place more emphasis on development of sustaining foundations for dealing with complex resource issues. Learning is a long-term proposition that requires a ballast against short-term politics and objectives. Another shift will likely require a change in the focus of actions away from management by objectives and determination of optimum policies and toward new ways to define, understand, and manage these systems in an ever-changing world. That focus should not be solely on variables of the moment (water levels, population numbers) and their correlative rates but rather on more-enduring system properties such as resilience, adaptive capacity, and renewal capability. This framework involves both the human components of the system (operations, rules, policies, and laws) and the biophysical components of the landscape and its ecosystems. The shift of focus to a learning basis is likely to require flexible linkages with a broader set of actors or a network.

Literature Cited

Ball, M. M., E. A. Shinn, and K. W. Stockman. 1967. The geologic effects of Hurricane Donna in South Florida. *Journal of Geology* 75:583–597.

Bosence, D. 1989. Biogenic carbonate production in Florida Bay. *Bulletin of Marine Sciences* 44:419–433.

Carlson, P. R., L. A. Yarbro, and T. R. Barber. 1994. Relationship of sediment sulfide mortality of *Thalassia testudinum* in Florida Bay. *Bulletin of Marine Science* 54:733–746.

Cohen, A. D. 1984. Evidence of fires in the ancient Everglades and coastal swamps of southern Florida. In *Environments of South Florida: Present and Past.* Memoir 2. Edited by P. J. Gleason. Coral Gables, Fla.: Miami Geological Society.

Craighead, F. C., Sr. 1971. *The trees of South Florida.* Vol. 1, *The natural environments and their succession.* Coral Gables, Fla.: University of Miami Press.

Davis, S. M. 1989. Sawgrass and cattail production in relation to nutrient supply in the Everglades. In *Fresh water wetlands and wildlife,* edited by R. R. Sharitz and J. W. Gibbons. Department of Energy Symposium Series No. 61. U.S. Department of Energy, Office of Scientific and Technical Information. Oak Ridge, Tenn.

———. 1994. Phosphorus inputs and vegetation sensitivity in the Everglades. Pp. 357–378 in *Everglades, the ecosystem and its restoration,* edited by S. M. Davis and J. C. Ogden. Delray Beach, Fla: St. Lucie Press.

Davis, S. M., and J. C. Ogden.1994. Towards ecosystem restoration. Pp. 769–796 in *Everglades, the ecosystem and its restoration,* edited by S. M. Davis and J. C. Ogden. Delray Beach, Fla: St. Lucie Press.

Duever, M. J., J. F. Meeder, L. C. Meeder, and J. M. McCollom. 1994. The climate of south Florida and its role in shaping the Everglades ecosystem. Pp. 225–248 in *Everglades, the ecosystem and its restoration,* edited by S. M. Davis and J. C. Ogden. Delray Beach, Fla: St. Lucie Press.

Durako, M., and K. Kuss. 1994. Effects of *Labyrinthula* infection on the photosynthetic capacity of *Thalassia testudiunum. Bulletin of Marine Science* 54:722–732.

Ehrlich, P., and A. Ehrlich. 1993. *The betrayal of science and reason.* Washington, D.C: Island Press.

Fourqurean, J. W., R. D. Jones, and J. C. Zieman. 1993. Processes influencing water column nutrient characteristics and phosphorus limitation of phytoplankton biomass in Florida Bay, Florida, USA: Inferences from spatial distributions. *Estuarine, Coastal and Shelf Science* 36:295–314.

Ginsberg, R. N., and H. A. Lowenstam. 1958. The influence of marine bottom communities on the depositional environment of sediments. *Journal of Geology* 66:310–318.

Gunderson, L. H. 1992. Spatial and temporal hierarchies in the Everglades ecosystem. Ph.D. diss., University of Florida, Gainesville.

———. 1994. Vegetation: Determinants of composition. Pp. 323–340 in *Everglades, the ecosystem and its restoration*, edited by S. M. Davis and J. C. Ogden. Delray Beach, Fla.: St. Lucie Press.

Gunderson, L. H., and W. F. Loftus. 1993. The Everglades. Pp. 199–255 in *Biotic communities of the southeastern United States*, edited by W. H. Martin and A. C. Echternacht. New York: John Wiley and Sons.

Herndon, A. K., L. H. Gunderson, and J. R. Stenberg. 1991. Sawgrass, *Cladium jamaicense*, survival in a regime of fire and flooding. *Wetlands* 11:17–27.

Holling, C. S. 1973. Resilience and stability of ecological systems. *Annual Review of Ecology and Systematics* 4:1–23.

Loveless, C. M. 1959. A study of the vegetation of the Florida Everglades. *Ecology* 40:1–9.

Newman, S., J. B. Grace, and J. W. Koebel. 1997. Effects of nutrients and hydroperiod on *Typha*, *Cladium*, and *Eleocharis*: Implications for Everglades restoration. *Ecological Applications* 7:1016–1024.

Newman, S., J. Schuette, J. B. Grace, K. Rutchey, T. Fontaine, K. R. Reddy, and M. Pietrucha. 1998. Factors influencing cattail abundance in the northern Everglades. *Aquatic Botany* 60:265–280.

Perkins, R. D., and P. Enos. 1968. Hurricane Betsy in the Florida-Bahama area: Geologic effects and comparison with Hurricane Donna. *Journal of Geology* 76:710–717.

Robblee, M. B., T. R. Barber, J. P. R. Carlson, M. J. Durako, J. W. Fourqurean, L. K. Muehlstein, D. Porter, L. A. Yarbro, R. T. Zieman, and J. C. Zieman. 1991. Mass mortality of the tropical seagrass *Thalassia testudinum* in Florida Bay (USA). *Marine Ecology Progress Series* 71:297–299.

Steward, K. K., and W. H. Ornes. 1975. The autecology of sawgrass in the Florida Everglades. *Ecology* 56:162–171.

Swart, P. K., G. F. Healy, R. E. Dodge, P. Kramer, J. H. Hudson, R. B. Halley, and M. B. Robblee. 1996. The stable oxygen and carbon isotopic record from a coral growing in Florida Bay: A 160 year record of climatic and anthropogenic influence. *Palaeogeography Palaeoclimatology Palaeoecology* 123:219–237.

Tabb, D. C., D. L. Dubrow, and R. B. Manning. 1962. *The ecology of northern Florida Bay and adjacent estuaries*. State of Florida Board of Conservation, Technical Series No. 39. Tallahassee, Fla.

Thayer, G. W., P. L. Murphey, and M. W. LaCroix. 1994. Responses of plant communities in western Florida Bay to the die-off of seagrasses. *Bulletin of Marine Science* 54:718–726.

Tilmant, J. T., R. W. Curry, R. Jones, A. Szmant, J. C. Zieman, M. Flora, M. B. Robblee, D. Smith, R. W. Snow, and H. Wanless. 1994. Hurricane Andrew—effects on marine resources. *BioScience* 44(4):230–237.

Wade, D. D., J. J.Ewel, and R. Hofstetter. 1980. *Fire in south Florida ecosystems*. USDA, Forest Service, General Technical Report SE-17, Southeast Forest Experimental Station, Asheville, N.C.

Walters, C. J. 1986. *Adaptive management of renewable resources*. New York: McGraw Hill.

———. 1997. Challenges for adaptive management of riparian and coastal ecosystems. *Conservation Ecology* [online] 1(2):1. http:www.consecol.org./vol1/iss2/art1.

Walters, C. J., L. H. Gunderson, and C. S. Holling. 1992. Experimental policies for water management in the Everglades. *Ecological Applications* 2:189–202.

Westley, F. 1995. Governing design: The management of social systems and ecological management. Pp. 391–427 in *Barriers and bridges to the renewal of ecosystems and institutions,* edited by L. H. Gunderson, C. S. Holling, and S. S. Light. New York: Columbia University Press.

Zieman, J. C., J. W. Fourqurean, and R. L. Iverson. 1989. Distribution, abundance and productivity of seagrasses and macroalgae in Florida Bay. *Bulletin of Marine Science* 44:292–311.

7

Ecological Resilience in Grazed Rangelands: A Generic Case Study

Brian H. Walker

Arid and semi-arid rangelands in which some form of pastoralism is the primary form of land use occur on all continents barring Antarctica, and collectively they are host to millions of people and livestock, and a very large array of biota: The Old World rangelands have a very long history of use, and their present structure and composition have been sculpted over many hundreds, and in some cases thousands, of years of pastoral activities. But it is over the past one hundred or so years that many of them, in Africa, North and South America, Australia, and parts of central and south Asia, have experienced major change. Increasing degradation seems to have accompanied the advances in technology in animal husbandry and water development.

This overview is restricted to the semi-arid and arid rangelands where regular, arable agriculture is precluded by insufficient rainfall. Subsistence farmers in these regions attempt to cultivate crops (sorghum and millet, and sometimes maize) in promising rainy seasons, but the mainstay for survival and commercial production is livestock and (under subsistence conditions) harvesting the native biota.

The Rangeland System, Structure, and Composition

The rangelands of the world are mostly characterized by a strongly seasonal climate (hot rainy season, cool dry season) with a high annual coefficient of rainfall variation, up to 40 percent or more (Huntley and Walker 1982). The soils are variable, with high spatial variation, but there is a major dichotomy in rangeland type, and in the ways in which rangelands respond to use, between the sandy rangelands and those on heavy soils.

The essential variables in the rangelands that are significant in terms of sustainable livestock production are woody plants (W—including shrubs and trees), grasses (G), which are divided into perennial grasses (P) and annual grasses (A), and the herbivores, which are predominantly domestic livestock (L), consisting of grazers (cattle and sheep) and browsers or mixed feeders (B) (such as goats and camels). In some areas there is also a significant biomass of wildlife.

An important feature of rangelands, characteristic of ecosystems where rainfall is insufficient to maintain a complete vegetative cover, is the spatial pattern resulting from redistribution of water. As described by Pickup (1985) and Tongway and Ludwig (1989), exposed soil surfaces tend to form a seal, decreasing the rate of water infiltration and thereby leading to increased run-off. The effect is a spatially heterogeneous pattern of soil conditions and associated plant communities, often in the form of a banded vegetation pattern. Water flowing across a run-off zone accumulates and eventually begins to infiltrate where it is slowed, initially by some surface obstruction. The increased soil water in this zone leads to development of a complete grass sward, which feeds back positively on infiltration (an order of magnitude higher; Kelly and Walker 1976; Greene 1992), and this induced run-on zone captures the water and topsoil flowing across the run-off zone above. Where this effect runs out, the vegetative cover thins, and run-off starts again. The net effect of this spatial pattern is the existence of patches of high productivity consisting of perennial grasses and sometimes woody plants, and a background mosaic of much less productive vegetation. The significance of this run-off/run-on patterned landscape is dependent on soil type. Sandy soils have inherently high infiltration rates and do not form soil surface seals, and accordingly do not exhibit this kind of spatial patterning, though a similar effect can be induced by wind erosion, with the formation of dunes and swales.

Rangeland Dynamics

This section details the dynamic properties of rangeland ecosystems, with emphasis on the external variables that drive the system (rainfall, fire, grazing and browsing), the dynamic ecosystem processes (e.g., competition, succession, fuel accumulation, and their interaction with drivers), and the ecosystem states through which the system moves.

The Drivers

Rainfall.   A key characteristic of rainfall in rangelands is high interannual variation, leading to high interannual variability in fodder production. Rainfall is also strongly seasonal, leading to an intra-annual fluctuation in the amount and quality of herbage.

The ratio of woody plants to grasses in most rangelands is a function of rainfall and soil type. For any particular rainfall, sandy soils support a higher proportion of woody plants than do clay soils, because more of the incoming rainfall is available for plant growth (lower wilting point) and because more of the rainfall gets through to the sub-soil, where woody plants have a competitive advantage over grasses (Knoop and Walker 1985). The dynamics of woody plants and grasses are described in the following section.

Fire. The frequency and intensity of fire depends on fuel load, and fuel load is predominantly determined by accumulated perennial grass. Annual grasses break down and the dead biomass does not accumulate. Woody vegetation contributes very little to fuel load. Fire kills woody seedlings, if sufficiently intense, and sets back developing woody plants. The net effect is a more open rangeland with much less of a shrub layer than occurs in the absence of fire.

Grazing and browsing. Cattle are the predominant domestic livestock in tropical rangelands; cattle and sheep are of equal significance in temperate regions. In rangelands under subsistence use, goats and sometimes camels can be almost as abundant, with native ungulates making a small contribution. In wildlife regions, grazers and browsers are generally in proportion to the available grass and browse.

The Dynamics

In their natural state, woody plants and grasses compete for soil water, and woody plants are the superior competitor. Within the grasses, perennial grass (P) is dominant over annual grass (A).

High interannual variation in rainfall keeps the amount of woody plants lower than the amount that the average rainfall could sustain. This is because dieback of woody plants is very rapid in a drought, but recovery during wet conditions is much slower, limited by both demographic and physiological processes. The amount of woody vegetation is therefore determined by the dry years. This allows grass to remain in the system at a significant level, using the water that woody vegetation is unable to take up. If grazing is light, accumulating perennial grasses (P) allow for periodic fires that kill small woody plants and set back established woody plants (W). This leads to an even higher $P:W$ ratio than the average rainfall would allow and results in the characteristic unstable mix of grasses and woody plants fluctuating over time in response to variation in rainfall and periodic fires.

A vigorous perennial grass layer prevents establishment of woody seedlings and also prevents soil erosion. High grazing intensity reduces perennial grasses, allowing, first, unpalatable perennial grasses (UP) and then annual grasses to take over. Annuals (which have weaker root systems) do not control woody seedlings, do not accumulate a fuel load, and are less effective in controlling soil erosion.

Fire frequency is a function of perennial grass production (and therefore rainfall) and the amount of grazing. In natural wildlife systems it is characteristically around three to ten years for savannas ranging from around 1,000 millimeters down to around 400 millimeters of rainfall per year.

When woody plants establish under high livestock pressure, they generally do so in a wet period, resulting in a cohort of seedlings that progress through to an even-aged, dense stand that exerts a much stronger competitive effect on the grass layer than does an age-structured stand of large and smaller trees and shrubs. If this developing cohort increases beyond some critical level of woody plants, then there is insufficient perennial grass for an effective fire. The system then continues to increase in woody vegetation up to some maximum amount, in the form of a "thicket," and stays in this state until individual woody plants begin to die (from old age) and the restructuring of the vegetation allows grasses to come back into the system. As perennial grass increases, fire again enters the system.

A change from perennial to annual grasses through heavy grazing results in lower grass production, much greater interannual variation, and increased soil erosion.

The States

In the lake ecosystem case study, Carpenter and Cottingham (1997; see chapter 3) describe lakes as being in either a "normal" or a "pathological" state. The normal state, dominated by fish and with few algae or macrophytes, is the preferred state as far as lake users and managers are concerned. The pathological state is dominated by high phosphorous content from run-off due to surrounding land use, algal blooms and aquatic weeds with few fish, and from the lake manager's viewpoint is undesirable. Some of the surrounding land managers may also consider this to be the case, but it is an externality as far as they are concerned and doesn't enter into their economic calculations of costs and benefits. In the rangelands case, the analogous normal and pathological states are the grasses (perennial)–woody plants mix and the woody plant-dominated states, respectively. The difference is that in the rangelands case the change in state is not externally driven. It does, however, depend on an interaction between an external (rainfall) and an internal driver (stocking rate), but the outcome is largely under the control of the manager.

Rangeland Resilience: Mechanisms and Issues

Any discussion of resilience in a particular ecosystem must be prefaced by the question, "The resilience of what to what?" The term resilience is used in two ways: (1) the amount of change the system can endure in the face of external shocks without changing "state" (Holling 1973; Walker et al. 1981), and (2) the rate at which it returns to its equilibrium "state" after disturbance (the "characteristic return time," Pimm 1982).

The first focuses on the behavior of the system when far from equilibrium—how much it can change and still recover; the second, which is measured by the largest (least negative) eigenvalue of the interaction matrix, focuses on the behavior of the system in the neighborhood of the equilibrium. In dealing with change in state (which is the focus of this analysis) it is the first definition that is of interest. The behavior near equilibrium is more of academic interest than it is of value to management.

To become operational, the system needs to be defined in terms of (1) the variables that describe the state, and (2) the nature and measures of the external shocks. In rangelands management, the "state" that is of interest is sometimes simply grass biomass. At other times (e.g., in a national park) biomass may be unimportant, and the "state" is defined by the proportional composition, or even just the presence, of a particular set of species. The two may not be unrelated (see below) but it is necessary to clearly identify which aspect of system structure is of concern. In what follows I take the "state" of a rangeland to be defined by (1) its productive potential (the amount of plant biomass produced per millimeter of rainfall, taken over the rangeland as a whole—in other words, including all spatial variation), and (2) the proportional composition of the plant biomass in terms of woody plants, perennial grasses, unpalatable perennial grasses and annual grasses. There is no reason to expect a system to be generally resilient, in other words, resilient to all forms of disturbance, and in rangelands it is resilience in the face of rainfall fluctuations, fire, and grazing that is of prime concern. Furthermore, the evidence suggests that most changes in state occur in response to the interactive effects of two or more simultaneous external shocks, for example drought and heavy grazing (Hodgkinson 1996; Walker and Noy-Meir 1982).

The Distinction between Sandy and Heavy Textured Soils

Sandy soil rangelands are inherently more resilient than those with lots of silt and clay, for three main reasons:

- They have a large underground biomass, which is unavailable to livestock and fire.
- They have high infiltration, low run-off and therefore little or no soil erosion, even when bared of vegetation.
- They are infertile, and the vegetation they support consequently has a very high C:N ratio which makes it relatively unpalatable, and it therefore has a high proportion which is largely inedible. An extreme example is the *Spinifex* grasslands of central Australia.

Species Diversity and Resilience

The performance of ecosystems and the likelihood that they will persist in terms of the services they provide is best discussed in terms of functional types of organisms, rather than just species. As a general statement, the functions a rangeland (or any other ecosystem) performs depends on the continued performance of a critical set of functional types (FTs). Diversity of FTs is therefore necessary for continued ecosystem function. Diversity within FTs confers resilience on ecosystems, in that a range of species with different environmental responses and capacities to withstand different stresses allows for compensatory behavior in terms of the function they perform. It is much safer for a rangeland to have several nitrogen-fixing plants species than just one.

Diversity of species also confers stability of function (where stability is taken to mean constancy over time), as illustrated by McNaughton's (1985) work in the Serengeti. Those grasslands with high numbers of grass species varied less in terms of biomass than those with fewer species.

Maintaining Resilience through Disturbance

The notion that the boundaries of resilience with respect to grazing pressure are maintained by periodic heavy grazing was described in Walker et al. (1981)

using examples and a model from southern Africa. In the face of grazing pressure, resilience of biomass and grass cover is conferred by the presence of unpalatable grazing-resistant species. Such species are out-competed over time by palatable species with higher relative growth rates, and it is only through periodic bouts of heavy grazing pressure that a balance of palatable perennial grasses and unpalatable perennial grasses is maintained.

Ludwig et al. (1997) introduced the concept of stability and resilience in ecosystems using the simple analogy of a floating raft. They postulate a raft with people and a weight located somewhere on it. The magnitude of the weight determines the boundaries of movements of the people around the raft that prevent the raft from capsizing (and changing "state"). The smaller the weight the more people can safely move around (i.e., the more resilient is the raft "system"). Suppose now that unless the people move around near the limits of safety, the weight inexorably increases in size. Only by getting the raft to wobble a bit can some of the weight be dislodged and lost overboard. Keeping everyone near the fulcrum may keep the raft stable in the short term, but it leads to a constant decline in the safe limits to movement around the raft. If we add this extra dimension to the Ludwig et al. analogy, it captures the necessary ingredients for reflecting the resilience of the savanna rangelands.

Returning to the real rangelands, constant heavy grazing leads to the death of perennial grasses (P) and replacement by unpalatable perennials (UP). Recent work in northern Australian rangelands (Walker et al. 1997) shows that after a disturbance leading to dominance by UP, recovery of the original $P{:}UP$ composition will depend on management during subsequent episodic rainfall events, when drought (leading to death of established UP) is followed by rainfalls that enable reestablishment of perennial grasses in the space provided. This pattern was also clearly identified for the Mitchell grass plains of Queensland by Austin and Williams (1988), who showed that such events were linked to the El Niño–La Niña rainfall sequences.

These dynamics highlight an important aspect of rangelands resilience for management. There is a hysteresis effect in the recovery toward a higher $P{:}UP$ composition as livestock pressure is reduced, compared to the levels of livestock pressure that brought about the initial reduction in the $P{:}UP$ ratio (Walker 1993).

Rangeland Dynamics near the Boundary of Resilience: The Issue of Time Scales

I consider here the change in state from a rangeland consisting of a mix of woody plants and grasses to one dominated by woody plants (a thicket state) (Walker et al. 1982; Ludwig et al. 1997; Ludwig et al. chapter 3).

The change from a mixed grasses–woody plants system (with 50 percent of production as grasses) to the woody plants thicket state is at first sight an irreversible transition—a change that would correspond to a transition across the boundary from one domain of attraction to another. A reversal can be brought about by a fire that reduces the woody vegetation, but such a fire is possible only if there is sufficient fuel present. Under continuous high grazing pressure there is a gradual increase in W, and beyond some critical level, even if all livestock are removed, there is insufficient P to allow for a fire. The return to the normal state is then a very slow process (forty to fifty years). Hence there may be two domains of attraction if the system is considered over a relatively short time scale but only a single domain of attraction if considered over a longer period. The difference may be unimportant for practical purposes, since farmers cannot afford to wait for fifty years for a return to favorable conditions.

Where perennial grasses have been killed by high grazing pressure coupled with drought, recovery of perennial grasses from an annual grass sward requires removal of livestock and above-average rainfall.

If the system stays in a state dominated by annuals, soil erosion occurs. Soil erosion is the net loss of soil from the system, rather than small-scale redistribution, and represents an irreversible change to a lower production state.

The Effect of Patchiness

As described earlier, on heavy textured soils in regions where rainfall is too low for complete vegetation cover, spatial patterning strongly influences both the overall species composition (with increased species diversity) and net production. Loss of this pattern, or an increase in the extent of the patches, leads to increased run-off surfaces and often to an increase in (or dominance by) woody plants in the run-on areas, with increased net loss of soil and therefore a reduction in overall productivity.

A final component of resilience is one that was conferred on rangelands in their natural states by large-scale movements of herbivores—the migratory behavior of the large grazing herds of Africa and North America. Such migrations have now all but ceased, except in a few remaining wildlife regions. Fryxell et al. (1988) have proposed a mechanism, and proffered evidence, to show that in regions where there is a strong seasonal variation between sub-regions, overall regional productivity and sustainable levels of herbivore biomass is higher where migration occurs, compared to the same region with fixed spatial locations of herbivores.

A postscript comment on resilience issues and mechanisms in rangelands is that virtually all grazed rangelands exist as an unstable mixture of W and G.

Managers both of livestock and wildlife don't want the rangeland to be at an equilibrium state, even if it were feasible to achieve it.

Indicators and Measures of Resilience

It is difficult to identify precise measures of resilience. As discussed earlier, the measures have of necessity to be related to particular kinds of change. The issue of whether there are general properties of ecosystems that render them resilient in the face of all external pressures or shocks needs to be further developed. From what is known about savanna rangeland dynamics two candidate measures that might indicate change in resilience are (1) the ratios of $W:P:A$, and (2) the scale and pattern of patchiness.

Regarding the former, increasing $W:P$ indicates loss of resilience in regard to the maintenance of the mixed grasses–woody plants system. Decreasing $P:UP$, or $P:A$ indicates a change toward a less-desirable state, but it is not clear if particular changes in the ratios correspond to loss of resilience (where resilience in this case is taken to be capacity to recover edible forage production), because a certain amount of unpalatable perennial grass confers resistance to grazing pressure by maintaining a vegetative cover on the soil.

Regarding the scale and pattern of patchiness, decreasing spatial heterogeneity indicates declining resilience—more extensive run-off and a greater likelihood of net loss of soil under heavy rainfalls.

Concluding Remarks

Most undesirable changes in rangelands are due to lack of understanding of the interactive effects of the drivers of rangelands dynamics and the fact that there are threshold effects in these dynamics that result in irreversible or difficult-to-recover-from changes in composition and production.

The effects of spatial differentiation are an important contributor to both overall productivity and the resilience of the system to variation in rainfall.

There is a limit to grazing intensity and this limit is usually surpassed when high stocking rates and a drought coincide. In many countries, governments tend to subsidize grazing during such times in the form of "drought relief" programs, leading to unwanted changes in state. An analogous situation occurs in wildlife areas when managers provide extra water during droughts, preventing animal mortality and prolonging excessive grazing pressure. Institutional drivers in these cases reduce resilience.

Some effects are counterintuitive, such as the need to periodically graze heavily in order to maintain the species that enable the rangeland to withstand heavy grazing without changing state. Savanna rangelands naturally exist as unstable mixtures of trees, shrubs, palatable perennial grasses, unpalatable grasses, and annual grasses, constantly tracking an ever-changing equilibrium combination. Fluctuations in their proportions confer resilience on the rangeland in terms of both the persistence of the species involved and the level of available graze. Attempting to fix the composition in some particular (perceived as desirable) combination reduces the ability of the rangeland to continue fluctuating in a state space that includes desirable amounts of grass.

The change in state from a mix of grasses and woody plants to a woody thicket corresponds to a soft stability change (Ludwig et al. 1997; Ludwig et al. chapter 3), albeit that the reversal may take many decades. Viewed on a decadal (management) time scale such a change conforms to a change in state. For some of this time the system is actually in a different stability domain, attracted toward the thicket condition. But at some point the system inevitably changes such that it is attracted toward the former mixed grasses–woody plants state, though the dynamics are very slow. To effect a reversal to the former state at this time, on a management time scale, still requires mechanical intervention. The system is effectively, for a decade or more, in a different state. Viewed on a multi-decadal time scale the change conforms to one representing slow dynamics far from equilibrium but still within the same stability domain.

The change in state from a perennial grass-dominated to an unpalatable perennial grass-dominated rangeland conforms to a hard stability change (Ludwig et al. 1997) in so far as it involves a hysteresis effect. On a decadal time frame, the stocking density required to effect a reversal from the unpalatable perennial grass to the perennial grass state is very much lower than the stocking density that brought about the change from perennial grasses to U. But once again, viewed over a multi-decadal time scale, a reversal will occur, slowly and in an episodic manner, at intermediate stocking densities, indicating that the internal dynamics of the system conform to a single stable state with very slow dynamics far from equilibrium. It is not possible (or sensible) to discuss resilience without specifying the time scale.

Literature Cited

Austin, M. P., and O. B. Williams. 1988. Influence of climate and community composition on the population demography of pasture species in semi-arid Australia. *Vegetatio* 77:43–49.

Carpenter, S. R., and K. L. Cottingham. 1997. Resilience and restoration of lakes. *Conservation Ecology* [online]1(1):2. http://www.consecol.org/vol1/iss1/art2.

Fryxell, J. M., J. Greever, and A. R. E. Sinclair. 1988. Why are migratory ungulates so abundant? *American Naturalist* 131:781–798.

Greene, R. S. B. 1992. Soil properties of three geomorphic zones in a semi-arid mulga woodland. *Australian Journal of Soil Research* 30:55–69.

Hodgkinson, K. 1996. A model for perennial grass mortality under grazing. Pp. 240–241 in *Proceedings of the 5th International Rangeland Congress*. Society for Range Management, USA.

Holling, C. S. 1973. Resilience and stability of ecological systems. *Annual Review of Ecology and Systematics* 4:1–23.

Huntley, B. J., and B. H. Walker, eds. 1982. *Ecology of tropical savannas*. Berlin: Springer-Verlag.

Kelly, R. D., and B. H. Walker. 1976. The effects of different forms of land use on the ecology of a semi-arid region in south-eastern Rhodesia. *Journal of Ecology* 64:553–576.

Knoop, W. T., and B. H. Walker. 1985. Interactions of woody and herbaceous vegetation in a southern African savanna. *Journal of Ecology* 73:235–253.

Ludwig, D., B. H. Walker, and C. S. Holling. 1997. Sustainability, stability and resilience. *Conservation Ecology* [online] 1(1):7. http://www.consecol.org/vol1/iss1/art7.

McNaughton, S. J. 1985. Ecology of a grazing ecosystem: The Serengeti. *Ecological Monographs* 55:259–294.

Pickup, G. 1985. The erosion cell—a geomorphic approach to landscape classification in range assessment. *Australian Rangelands Journal* 7:114–121.

Pimm, S. L. 1982. *Food webs*. New York: Chapman and Hall.

Tongway, D., and J. Ludwig. 1989. Vegetation and soil patterning in semi-arid mulga lands of Eastern Australia. *Australian Journal of Ecology* 14:263–268.

Walker B. H. 1993. Rangeland ecology: Understanding and managing change. *Ambio* 22:2–3.

Walker, B. H., J. L. Langridge, and F. McFarlane. 1997. Resilience of an Australian savanna grassland to selective and nonselective perturbations. *Australian Journal of Ecology*. 22:125–135.

Walker, B. H. D. Ludwig, C. S. Holling, and R. M. Peterman. 1981. Stability of semi-arid savanna grazing systems. *Journal of Ecology* 69:473–498.

Walker, B. H., and I. Noy-Meir. 1982. Aspects of the stability and resilience of savanna ecosystems. Pp. 556–590 in *Ecology of tropical savannas*, edited by B. J. Huntley and B. H. Walker. Berlin: Springer-Verlag.

8

Resilience of Tropical Wet and Dry Forests in Puerto Rico

Ariel E. Lugo, Frederick N. Scatena, Whendee Silver, Sandra Molina Colón, and Peter G. Murphy

Our objective was to develop a general framework for exploring the resilience of tropical forests for the eventual elucidation of general principles that may explain the similarities and differences that exist among forests from different life zones. We define resilience as the rate at which a forest stand recovers from large and infrequent disturbances (LIDs). Examples of LIDs are hurricanes, fires, droughts, landslides, and other events that disrupt significant areas of forests. There is insufficient knowledge to assign thresholds of intensity, frequency, and geographic extent to LIDs. Resilience is measured by the slope of the relationship between any parameter of forest structure or function plotted over time after a LID event. The units can be expressed as percent recovery per unit time.

Using examples from subtropical (*sensu* Holdridge 1967) dry and wet forests in Puerto Rico (these forests are within the tropical latitudes), we identify five criteria of the framework: (1) defining levels of a hierarchy used as the basis for comparisons; (2) describing and comparing the structure and functioning of mature stages of the two forest types selected for the case study, and summarizing results that illustrate how resilience issues might be approached; (3) discussing the interaction of disturbances with ecosystems; (4) describing disturbances and forest responses to them; and (5) identifying generalized resilience mechanisms.

We expect that some aspects of forest structure and functioning respond to latitudinal-level LIDs such as hurricanes and will be common to forests with contrasting life zone and edaphic conditions. Other aspects of forest structure

195

and functioning respond to local climate and edaphic conditions and are unique to particular forest types irrespective of latitudinal conditions, in other words, temperature, photoperiod, and seasonality. Interpretation of resilience properties might also require attention to level of complexity within a hierarchy.

Caribbean Forest Hierarchy

The two forest types that we selected occur in the same latitudinal belt (subtropical), overlap in elevation, and diverge at the life zone and plant association level. We will refer to one forest type as the wet forest. This forest type ranges in elevation from lowland to pre-montane, and it is in the perhumid humidity province and the subtropical wet forest life zone. The plant association is the tabonuco forest on volcanic soil with various successional stages from mature to human- or hurricane-disturbed. The other forest type will be designated as the dry forest. This is a lowland forest in the subhumid humidity province and the subtropical dry forest life zone. The plant association is the deciduous to semi-evergreen forest on calcareous substrate and various stages of succession from mature to human- or hurricane-disturbed. This hierarchical classification allows precise description of the conditions under which these particular tropical forests function (table 8.1).

Stands are located in the Caribbean island of Puerto Rico. The wet forest is on the north coast in the Luquillo Mountains while the dry forest is located on the south coast at Guánica. Their geographic setting is contrasting because the presence of the Luquillo Mountains influences the climate of the wet forest whereas the dry forest is on a lowland setting several kilometers from the mountains. Originally, both landscapes were approximately 100 percent forested and were deforested and heavily fragmented by human activity for over a century (Domínguez Cristóbal 1989a–f; Scatena 1989; García Montiel and Scatena 1994; Molina Colón 1998). The forces of deforestation and fragmentation originated from systems at larger scales of complexity, in other words, those powered by the agrarian economy of the island (Lugo 1996).

Subsequent to abandonment of human use, the landscape was naturally reforested by the coalescence of small forest fragments into continuous forest cover (Lugo et al. 1996; Thomlinson et al. 1996). The recovery of the landscape was driven by both small- and large-scale processes. At the larger scale are changes in the economy of the island that released pressure on land use and allowed succession to proceed toward a forested landscape. At the smaller scale is the function of birds, ants, and bats as long- and short-distance dispersal agents for propagules from forested areas located at various distances from forest patches and abandoned agricultural lands. Today, the forested landscape is a

Table 8.1. Environmental conditions at the life zone and site levels of complexity

Parameter	Subtropical Wet Forest	Subtropical Dry Forest
FOR THE LIFE ZONE		
Mean annual rainfall (mm)	2,000 to 4,000	500 to 1,000
Mean annual biotemperature[a] (°C)	18 to 24	18 to 24
Ratio of potential evapotranspiration to precipitation	0.25 to 0.5	1.0 to 2.0
Elevation (m)	0 to 500	0 to 600
FOR SITES DISCUSSED HERE		
Mean annual rainfall (mm)	3,537	860
Mean annual biotemperature[a] (°C)	21–22	21–22
Ratio of potential evapotranspiration to precipitation	0.33	1.4
Elevation (m)	150 to 500	125 to 145
Main natural disturbance regime	Hurricanes	Hurricanes
Main anthropogenic disturbance	Deforestation/ Land use change	Deforestation/ Land use change

Sources: Data are from Holdridge 1967, Odum et al. 1970, Murphy and Lugo 1986ab, García-Martinó et al. 1996.
[a]Mean annual biotemperature is the mean of daily air temperatures with a value of 0 assigned to air temperatures above 30°C and below 0 °C.

mosaic of different-age stands that reflect past natural and anthropogenic disturbances. The forest types were selected in part because of their contrasting rainfall and edaphic conditions but shared biotemperature and disturbance regimes, and partly because of the availability of ecological data.

Overview of Forest Structure and Dynamics

In this section we describe, compare, and contrast the structure and dynamics of two types of forest: subtropical wet forest (also called *tabonuco forest*) and subtropical dry forest (containing a mixture of deciduous and semi-evergreen trees).

Subtropical Wet (Tabonuco) Forest

This forest is an evergreen forest with tall, straight-bole trees whose physiognomy creates a cathedral-like effect when viewed from the ground. The forest interior is dark as a result of a closed canopy. It has vertical plant stratification, many life forms, a few tree species with high importance value (Smith 1970), and dominance of tree size classes which reflects cohorts of different age classes (Lugo and Scatena 1995; Scatena and Lugo 1995). Fine litter stocks on the forest floor are low and variable in time and space (Lugo 1992; Lugo and Scatena 1995). Trees of one of the dominant species (*Dacryodes excelsa*) form unions, which are groups of trees connected by root grafts (Basnet et al. 1993). The majority of the root system of the tree unions is relatively shallow but some roots can penetrate one meter or more. These tree unions influence ecosystem function and resilience to hurricanes (Basnet et al. 1993). The canopy of this mature forest is smooth in appearance when viewed from above and lacks emergent trees because dominant individuals attain similar height (Odum 1970).

The functioning of the wet forest is characterized by high annual rates of primary productivity, high respiration rates, intermediate rates of evapotranspiration, high water interception, and high nutrient use-efficiency (Odum 1970; Scatena 1989; Lugo 1992; Silver 1994; Lugo and Scatena 1995; García-Martinó et al. 1996). Forest succession has several distinct stages and is fast, in other words, about sixty years from a small clearcut or a hurricane gap to characteristics of a mature stand (Lugo and Scatena 1995; Lugo et al. 1999).

Subtropical Dry (Deciduous to Semi-evergreen) Forest

This forest contains a mixture of deciduous and evergreen species of low stature, a high density of small-diameter trees, and poorly defined vertical plant stratification (Lugo et al. 1978; Murphy and Lugo 1986b). Trees with multiple stems and peeling bark are common (Murphy and Lugo 1986a,b). Root biomass is a high fraction of total stand biomass. Roots form a thick and shallow mat, but they can also penetrate at least a meter into the soil (Murphy and Lugo 1986b). A high mass of fine litter accumulates on the forest floor (Lugo et al. 1978; Lugo and Murphy 1986). The canopy is closed during the wet season without emergent trees, giving the appearance of a smooth and continuous green carpet over the land surface. In the dry season the canopy loses leaf area, and opens (Murphy et al. 1995). Tall trees lose their leaves first, giving the impression of a leafless canopy over a green understory. Herbaceous plants grow in the understory at the beginning of the rainy season before the canopy closes. Termites and subterranean ants are important ecosystem components as they accelerate organic matter, soil, and seed turnover. The avifauna is particularly diverse in comparison to that of the wet forest (Kepler and Kepler 1970).

The dry forest has moderate annual primary productivity, high respiration rates, a high ratio of potential evapotranspiration to precipitation, and low rates of evapotranspiration relative to the wet forest (722 versus 1,707 millimeters per year; Lugo et al. 1978; Lugo and Murphy 1986, García-Martinó et al. 1996). Nutrient use-efficiency is high, particularly for phosphorus. Plant succession after small-scale clearcutting is surprisingly fast and initially dependent on root and stem sprouting of vegetation (Ewel 1977; Murphy et al. 1995). Because of the abundant sprouting of preexisting individuals, dry forests have few distinct stages in the plant succession after mechanical disturbance.

Comparison of Wet and Dry Forests

In spite of their contrasting climatic and edaphic conditions, these two forest types share similar latitudinal and altitudinal conditions (table 8.1). They also share similar biotemperature and atmospheric disturbances, i.e., low pressure systems and hurricanes. Because of the ecological importance of biotemperature and disturbance regime, we anticipate that these two forest types might exhibit more ecological similarities when compared to each other than they do when compared with forests with similar rainfall or similar substrates but in different latitudinal and/or altitudinal locations, such as those wet or dry forests in tropical lowlands (sensu Holdridge 1967). For example, the biomass of epiphytes in the wet forest is closer to the value in the dry forest (table 8.2) than it is to the value in tropical moist forest (0.05 megagrams per hectare) near Manaus in the Amazon Basin (Klinge et al. 1975). We believe that the presence of trade winds in the Caribbean and their absence in the Amazon explain the biomass difference across, and its similarity within, latitudes (Lugo and Scatena 1992).

A mature wet forest has a greater structure than a mature dry forest: specifically, it has larger trees; more basal area, biomass, and wood volume; larger leaves and higher leaf area index (LAI); but lower average fine litter and soil organic matter stocks (table 8.2). In terms of processes, the wet forest has higher aboveground net primary productivity (defined as the sum of aboveground biomass increment plus total fine litterfall; total fine litterfall includes leaves, wood, flowers, fruits, and miscellaneous fall), litterfall, woodfall, and herbivory rates than does the dry forest. However, eight parameters resulted in similar values for these apparently contrasting forest types, including smooth canopies without emergent trees; high tree species dominance (sensu Whittaker 1965); tree species density (figure 8.1a); total tree species; annual leaf fall, wood volume, and wood biomass growth rates; and tree density for trees with diameter at breast height (dbh) greater than 4 centimeters.

We and others (Odum 1970; Lugo 1988, 1991; Lugo and Scatena 1995) interpret some of these similarities as resulting from the common latitudinal belt and trade-wind-dominated climate. Trade winds and hurricanes are probably

Table 8.2. Measures of structural and functional parameters of the mature study sites

Parameter	Subtropical wet	Subtropical dry
STRUCTURE		
Tree density (trees/ha)	1,750[a]	1,170 to 2,307[b]
Canopy height (m)	20–30	6–9
Common dbh range (cm)	4–50	3–8
Basal area (m²/ha)	40	7–11
Aboveground volume (m³/ha)	350	50[c]
Aboveground stemwood biomass (Mg/ha)	190	41
Epiphyte biomass (Mg/ha)	0.5	0.14
Leaf biomass (Mg/ha)	7.9	4
Leaf area index	6 to 7	2 to 4
Specific leaf area (cm²/g)	127	92
Total tree species	170[d]	169
Tree species/ha	50	52[e]
Importance Value of dominants (%)	19–30	10–21
Soil organic matter (Mg/ha)	146–160[f]	172–248[g]
Loose litter (Mg/ha)	6 to 8	9 to 15
Leaf litter (Mg/ha)	5 to 6[h]	7 to 12
PROCESSES		
Herbivory rate (Mg/ha·yr)	0.38	0.08[i]
Litterfall (Mg/ha·yr)	8.6–9.7	2.9–5.5
Leaf fall (Mg/ha·yr)	4.9–5.5	2.5–4.4
Woodfall (Mg/ha·yr)	1.4	0.4–0.8
Stemwood volume growth (m³/ha·yr)	2.26	2.33
Stemwood biomass growth (Mg/ha·yr)	2.5	2.1
Aboveground net primary productivity (Mg/ha·yr)	10.5	6.9
Litterfall/loose litter mass (yr⁻¹)	1.2	0.3
Leaf fall/leaf litter mass (yr⁻¹)	0.90	0.45
Stemwood growth/stemwood biomass (yr⁻¹)	0.01	0.05

Sources: Unless noted, data were modified from Weaver and Murphy 1990 for the wet forest, and Lugo et al. 1978, Lugo and Murphy 1986, Molina Colón 1998, Murphy and Lugo 1986b, and Murphy et al. 1995 for the dry forest.

[a]4 or more centimeters diameter at breast height.

[b]5 or more centimeters diameter at breast height.

[c]Estimated based on basal area and height.

[d]From Little et al. 1974.

[e]From Quigley 1994.

[f]Low value to 0.25 meter, Odum 1970; high value to 0.85 meter, Lugo 1992.

[g]Low value to 0.85-meter depth and high value to 0.15-meter depth. Shallow soil pits were on peat soils.

[h]From Wiegert 1970.

[i]From Benedict 1976.

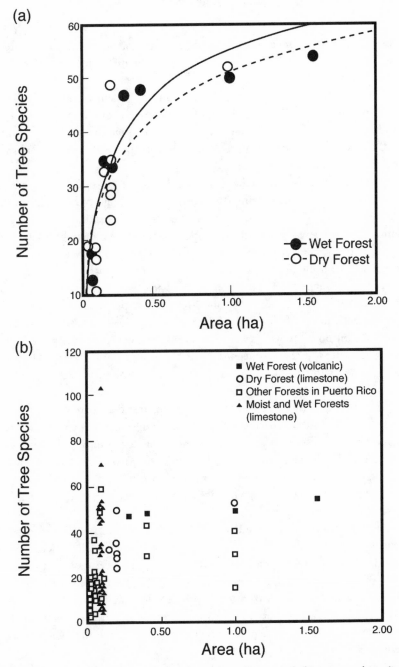

Figure 8.1. Species-area relations in *(a)* dry and wet forests and *(b)* various subtropical forests in Puerto Rico. Each point represents a different study. The logarithmic fit for the wet forest is $y = 28.8 \log(x) + 53.7$ ($r^2 = 0.84$), and for the dry forest is $y = 27.7 \log(x) + 50.6$ ($r^2 = 0.58$). Data sources are available from A. E. Lugo.

responsible for the similar upper canopy structure of these contrasting forests. Leaf area index and leaf size, however, adjust to the different humidity regimes of the forests.

High tree species dominance (table 8.2) and dominant size classes probably reflect the frequency, timing, and type of disturbances. For example, after a LID, conditions are optimal for the establishment of even-age cohorts of trees, which leads to the dominance of stands by a few species capable of rapidly establishing and growing after the disturbance.

Tree species density may reflect the synergy between disturbance regime and biotemperature of forests acting over geologic time. High biotemperature and geologic time favor the accumulation of species on islands, but LIDs favor the dominance of a few species with rapid post-disturbance growth responses. The eventual tree species density reflects the balance or synergy between these opposing trends, with the number of tree species per area decreasing with increasing level of disturbance. Most forests in Puerto Rico (with similar biotemperature and hurricane-dominated disturbance regimes) support similar number of species per unit area (figure 8.1b). The highest and greatest range of values are those of subtropical moist or wet forests on limestone substrate. In contrast to tree species density, the evergreen/deciduous species combinations reflect moisture regimes with deciduous species predominating toward the lower humidity levels (Chinea 1980).

Tree species density data were normalized for tree density to account for the different sizes of trees—specifically, large trees in the wet forest and smaller ones in the dry forest. Results show remarkable consistency in the number of species per thousand individuals across all forests in Puerto Rico (table 8.3) and support our suggestion that species richness is responding to overall conditions in the island, meaning the natural disturbance regime and geologic history, as opposed to variations in moisture.

At the landscape scale, the dry forest appears to have a higher tree species density than the wet forest does. Values for total tree species in table 8.2 correspond to about 4,000 hectares in the dry forest and to elevations between 150 and 600 meters in the wet forest (Little et al. 1974), or about 5,800 hectares (García-Martinó et al. 1996). However, these values are based on independent estimates that can change if both the species counts and area estimates are done in concert. Regardless, these results disallow the assumption that the tree species density is higher in the wet forest.

Similar rates of wood growth and leaf fall in dry and wet forests appear counterintuitive given the smaller wood volume and lower primary productivity of the dry forest. However, this comparison is for mature stands after they have experienced a successional growth pulse and reached relatively stable standing

Table 8.3. Tree species per thousand individuals (y´) based on regression equations between number of species per plot (y) and number of trees per plot (x) for which the r² and number of observations (n) are shown

Forest Type	Regression Equation	r^2	n	y'
Wet on volcanic	y = –43.78 + 30.89*LOG(x)	0.84	19	49
Moist and wet on limestone	y = –13.79 + 20.01*LOG(x)	0.37	39	46
Dry on limestone	y = –30.27 + 24.95*LOG(x)	0.52	26	45
Other forests in Puerto Rico	y = –20.75 + 20.46*LOG(x)	0.46	40	41
All forests	y = –21.85 + 22.16*LOG(x)	0.50	124	44

Source: Data include a variety of minimum diameters and are available from A. E. Lugo.

stocks of biomass. During the growth phase, a wet forest accumulates biomass at more than three times the rate of a dry forest (compare data in Scatena et al. [1996] and Fu et al. [1996] with data in Lugo et al. [1978] and Murphy and Lugo [1986b]) and reaches higher levels of structure. At maturity, the wet forest maintains a higher aboveground net primary productivity than the dry forest does because it produces more miscellaneous and fine woodfall. For this reason these forests can have similar leaf fall and stemwood growth rates and still be different in the rate of net primary productivity.

Moreover, the range of leaf fall in the wet forest is wider and more constant over time than the range in the dry forest, which tends to oscillate with rainfall. It is possible that with longer datasets it can be shown that the leaf fall rate of the wet forest oscillates at slightly higher values than that of the dry forest. The similarities in stemwood growth rate support the suggestion of Jordan (1971) that rate of wood production by trees is similar throughout the world. Nevertheless, the similar rates of leaf fall and stemwood production among these forest types influence the dynamics of these compartments. Because standing stocks of stemwood and leaf litter are different among the two forest types (table 8.2), the ratio of fluxes to standing stocks is different even if it has similar rate processes associated with it. For stemwood, this ratio is five times lower in wet forest than in dry forest, and for leaf litter it is twice the value in wet forest than in dry forest (table 8.2).

Similarities between dry and wet forests in tree density (for trees with a diameter at breast height greater than 4 centimeters) may be due to allometry or space constraints and not necessarily to any particular ecological factor. Both the wet and dry forests have high tree densities in small-diameter classes after disturbances, and both experience reduced tree densities as stands mature. However,

Table 8.4. Measures of nutrient use-efficiency in mature stands of subtropical dry and subtropical wet forests in Puerto Rico

Nutrient	Dry Forest	Wet Forest
NUTRIENT USE-EFFICIENCY OF LITTERFALL (MASS FALL/NUTRIENTS IN MASS FALL)		
N	98	92
P	6057	2999
RETRANSLOCATION[a] (KG/HA·YR)		
N	29.0	14.0
P	2.2	3.8
RETRANSLOCATION[a] (PERCENTAGE OF NEED[b])		
N	30	10
P	65	49
ABOVEGROUND NET PRIMARY PRODUCTIVITY/NUTRIENT UPTAKE		
N	123	98
P	6634	3154
K	160	384

Sources: Data are from Lugo and Murphy (1986) and Lugo (1992).
[a]Includes leaching.
[b]Nutrients required to satisfy annual net primary productivity.

at maturity, the higher tree densities in the dry forest can occur at the smaller size classes (less than 4 centimeters dbh) while mature wet forest has more trees at sizes greater than 4 centimeters dbh than at less than 4 centimeters dbh.

Table 8.4 compares wet and dry forests in terms of nutrient use-efficiency. The dry forest has higher phosphorus use-efficiency on all measures than the wet forest except for the absolute amount of phosphorus that is retranslocated. Nitrogen use-efficiency is slightly higher in the dry forest than in the wet forest, while the use-efficiency of potassium is higher in the wet forest than in the dry forest. Nutrient use-efficiency by these forests is probably more a reflection of local site conditions (substrate and moisture availability) than of larger scale phenomena. The dramatic difference in the efficiency of phosphorus use underscores this point.

The Interface between Disturbances and Ecosystems

Ecosystem stressors or disturbances are external forces that interact with particular interfaces of ecosystems (Lugo 1978). The point of interaction between the disturbance force and the ecosystem is the *biotic interface* (*sensu* Lugo and Scatena 1995; Silver et al. 1996 a, b). The organisms that occupy these interfaces are more exposed to the disturbance, absorb most of the impact, and can contribute to the recovery of the forest. For example, a wind storm will have its main direct effect on the forest canopy that is at the interface of the biotic system and the full force of the wind. Forest recovery after the event centers on the restoration of the canopy, a process that may last decades depending on post-disturbance site conditions and how much damage was caused by the wind storm.

Disturbance factors are usually present in ecosystems at a suite of intensities. At lower intensities organisms at the interface interact with these forces at levels that are below thresholds of damage. Wind for example, can range from breezes to gales and its effects range from beneficial to leaf gas exchange, to complete destruction of the stand. For this reason, the sources of resilience to members of a hierarchy in geographic space can be sought in at least five places:

1. In individuals as part of their responses to their environment.
2. In the cumulative effect of how organisms of different species react to their respective environments.
3. In the effect of legacies after an event, for example, surviving seedlings or root systems.
4. As the consequence of inputs from, or effects of, processes from other levels in the hierarchy under consideration.
5. As inherent characteristics of ecological systems, such as the negative feedback function of storages.

These resilience mechanisms may be adaptive (sensu Williams 1966) or fortuitous, but their contribution to ecosystem recovery can be shown through observation or experimentation. In our discussion below, we address responses from all five sources but make no claim that they are adaptive. However, we do argue that they contribute to forest resilience.

In reality, disturbances are complex events because they can involve more than one force, each with varying scales of intensity, frequency, duration, and area of influence. Such events have interactions with more than one biotic interface. An intense rainfall event, for example, interacts with the canopy where it can leach nutrients. It also interacts with wind and causes branch and leaf fall. The same event can interact with soils and topography to cause landslides and with soil and wind to cause tree falls. The consequences of these interactions (branchfall, leaf fall, landslides, and tree falls) are in themselves disturbance events with consequences to other sectors of the ecosystem, for example to understory plants or to microbial communities.

Wind and rain are the main disturbance forces associated with hurricanes

(Scatena and Larsen 1991). Each of these forces in turn has complex interactions with forest components such as the canopy, soil, roots, and animals. We simplify the analysis by first considering wind and water individually and later integrating the discussion of these two factors. For anthropogenic effects, we will consider ecosystem recovery after abandonment of different land uses. We take the precaution of identifying the land uses and some of the associated disturbance forces that may influence forest recovery, and we focus the discussion around five sets of questions.

Wind Storms

Caribbean forests are exposed to trade winds that blow in a northeasterly direction and regulate island climate. Puerto Rico is struck by a category 5 hurricane on an average of one per sixty years and by lower intensity storms at a greater frequency (Scatena and Larsen 1991). Every twelve years or so, the island experiences hurricane-force winds (Weaver 1987). Hurricane track records during the period 1871 to 1977 (Neumann et al. 1978) reveal spatial differences over the Caribbean region (from fewer than five to more than sixty hurricanes over particular locations), but most of Puerto Rico falls within the 42 (south coast) to 50 (north coast) hurricanes per 106-year isopleth. Over the geologic time that the flora of Puerto Rico has coexisted with hurricanes, this difference in the number of hurricanes between the north and south coast may not be significant, but their presence is clearly significant from the point of view of biotic responses and adaptation.

The first question that we will address is how Caribbean forests maintain resilience in light of the wide range of frequency and intensity of winds. One answer to this question is that the response of organisms to trade winds provides resistance to gale winds because the architecture of Caribbean forest canopies minimizes wind damage (Odum 1970; Brokaw and Grear 1991).

Lugo and Zimmerman (in press) examined the life cycle of trees subjected to LIDs and found the following characteristics: sprouting, formation of tree unions, small individuals, short life spans, rapid changes in sun and shade adaptation, rapid establishment of seedling populations, accelerated rates of primary productivity and nutrient cycling, and increased abundance of gap-dependent canopy species. The capacity to flower and fruit early in the life cycle or as a result of mechanical damage, (i.e., epicormic sprouting or stress flowering), is another resilience trait of trees in the Caribbean.

There are no published observations on the response of the dry forest to hurricanes. However, Hurricane Hortense passed over Guánica Forest as a category 3 hurricane in 1997 without causing appreciable structural damage to forest stands. Hurricane Gilbert passed in 1988 over dry forests in Jamaica and Quintana Roo, Mexico, and also caused little damage to stands (Whigham et al. 1991; Wunderle et al. 1992).

Canopy architecture, high dominance of tree species, and cohorts of similar size are hurricane-related responses that the dry forest shares with the wet forest. We also have detailed information on how the dry forest responds to experimental clearcuts (Ewel 1977; Murphy et al. 1995) and can compare its response with that of wet forests (Ewel 1977). We also include here the response of wet forests to hurricanes (Lugo and Scatena 1995; Scatena and Lugo 1995; Walker et al. 1991, 1996). Both types of disturbance—clearcuts and hurricanes—are examples of mechanical damage to aboveground structures of forests, although they each interact differently with their respective biotic interfaces.

The main initial effect of a wind storm such as Hurricane Hugo in 1989 is the mechanical destruction of structure, particularly the forest canopy (Brokaw and Grear 1991). This results in the massive transfer of biomass and nutrients to the forest floor (Lodge et al. 1991) and instantaneous mortality of organisms (Lugo and Scatena 1996). The first five years after the storm were characterized by high rates of delayed mortality (Walker 1995); delayed effects on other ecosystem sectors such as belowground carbon and nutrient pools (Parrotta and Lodge 1991; Silver et al. 1996c), and groundwater chemistry (McDowell et al. 1996); continuous high availability of soil nutrients throughout the period of recovery from the hurricane (Silver et al. 1996c); differential seedling responses to accumulated litter (Guzmán-Grajales and Walker 1991); and rapid rates of vegetation recovery (Scatena et al. 1996). A consequence of the damage caused by a wind storm is a dramatic change in the microenvironment, which sets the stage for ecosystem recovery and the possibility of species invasions.

Some trees respond to wind effects with regeneration by sprouting, increased seed rain, or both. The magnitude of the response depends on how extensive the damage was during the storm and whether individuals of the particular species have the ability to resprout or not. Massive germination of seed results in abnormally high seedling populations and a rapid turnover of plants as they compete for light and space (Scatena et al. 1996; Lugo and Zimmerman in press). As a result, there are short periods (about a decade) of high rates of turnover and structural change followed by longer periods (several decades) of subtle structural change (Lugo et al. 1999). The importance value of tree species can change and there are opportunities for species turnover and invasions at the patch scale of 1 hectare (Crow 1980; Lugo and Scatena 1995; Fu et al. 1996; Scatena et al. 1996).

After small-scale clearcutting, the recovery of canopy cover and leaf area index is fast in dry forests as is the recovery of leaf fall and nutrient return in leaf fall in

Table 8.5. Recovery of structural and functional parameters in subtropical dry and subtropical wet forests after clearcuts and hurricane damage (wet forest at age five years for leaf fall and nutrients in leaf fall)

Parameter	Age (yr)	Dry Forest	Wet Forest
STRUCTURE			
Leaf area index	8	90	
Cover	8	93	
Tree height	0.25	4	2
	0.5	6	4
	1.1	14	10
	5.5	34	
	11.0		44
	12		38
PROCESS			
Leaf fall	2	28	
	5		92
Nutrients in leaf fall	2	30	
	5		80

Sources: Data are from Murphy and Lugo (1986b), Lugo and Murphy (1986), and Scatena et al. (1996).

Note: Values are in percent of the control (dry forest) or original stands (wet forest). Age refers to time since the disturbance. An empty cell indicates no available data.

the wet forest (table 8.5). Using short-term data on recovery of tree height after a clearcut, Ewel (1977) concluded that the dry forest was more resilient than the wet forest (table 8.5). However, as we will show below, when forest response to other disturbances are evaluated with more ecological parameters over longer time periods, it is impossible to conclude that one forest is more or less resilient than the other. Both appear to be resilient to conditions of disturbance.

Water Availability

Annual rainfall in the Caribbean is cyclical. Some decades are relatively wet with rainfall greater than long-term averages, while other decades are relatively dry, with rainfall below long-term averages (figure 8.2). Long-term rainfall records show that while the two forests may receive different average rainfall, both experience similar variation and trends in rainfall. A second set of questions would be: How do Caribbean forests cope with such a wide range of rainfall conditions, that is, how resilient are they to widely varying rainfall regimes?

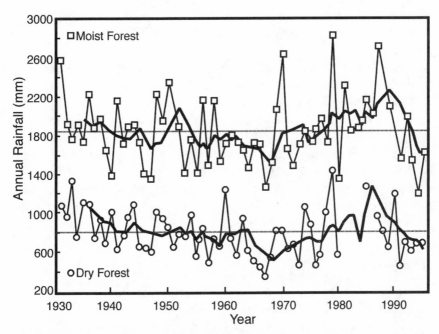

Figure 8.2. Long-term rainfall rates in moist and dry climate stations in Puerto Rico. The moist station corresponds to Canóvanas on the northern coast of Puerto Rico and the dry station to Ensenada on the southern coast. Long-term data are not available for a wet climate station, but available information suggests the temporal patterns are the same as these but with a higher annual rate of rainfall (García-Martinó et al. 1996). The heavy line is a five-year moving average and is based on data from NOAA.

Excessive Water

In the wet climate, high rainfall causes high rates of nutrient leaching (Lugo and Scatena 1995). On swales and slopes with shallow soils, high rainfall can also saturate the soil and reduce soil oxygen concentrations (Silver et al. 1999). Soils on slopes and valleys become unstable and experience mass movements and/or landslides after heavy rains (Scatena and Lugo 1995). Soil nutrients become unavailable to vegetation in high rainfall areas due to loss of topsoil horizons after mass movement or landslides, or reduced plant uptake capacity as a result of low soil oxygen concentration.

Excessive water is stored in soils, bromeliads, and coarse woody debris, leading to an abundance of microaquatic communities at these locations. There is also an increased dependence on insects and birds for plant pollination. Growth in wet environments is continuous and trees lack growth rings. Odum (1970) observed synchronization of many of the ecological processes of the wet forest to patterns in rainfall and changing photoperiod. Peaks of leaf fall and germination occurred with increasing rainfall and light availability after the lower-rainfall months.

Trees that grow on saturated soils have shallow root systems and thick root mats that contribute to recycling efficiency. The formation of tree unions with their conspicuous organic benches has the effect of isolating the tabonuco stands from mineral soil, facilitating soil aeration, and increasing internal nutrient recycling (Basnet et al. 1993). Plants in wet climates exhibit anatomical and morphological responses by the production of aerenchyma tissue and specialized gas exchange structures such as pneumatophores and lenticels. These tissues facilitate gas exchange in low soil oxygen environments. Coupled with trade wind effects, there is also an increase in epiphyte abundance that further contributes to nutrient sequestration and recycling.

Drought

Short-term drought may cause leaf overheating, tissue dehydration, and loss of leaf area index. As a result, plant growth is reduced or arrested. Plant uptake and nutrient transport to leaves decreases if transpiration ceases or is reduced drastically. Prolonged drought leads to progressive tree mortality, starting with top branches and individual trees and eventually killing large groups of trees.

Plants respond to drought with high tissue osmotic potential and anatomical and morphological structures that include leaf sclerophylly, water storage in stems, succulence, leaf movement and shedding, deep roots, and large root-to-shoot biomass ratios. Gas exchange processes such as respiration, photosynthesis, and transpiration are sensitive to humidity. Dry climates select plants with contrasting physiological properties, deep-rooted C_3 plants, drought-tolerant C_4 plants, such as grasses and other pioneer species, and CAM plants with succulent organs and nocturnal CO_2 fixation, such as cacti, agaves, bromeliads, and orchids. We do not have comparative information on the response of wet forests to drought, although droughts occur periodically in the wet forest life zones of Puerto Rico.

Land Use Conversion

Caribbean islands have high human population densities that range from ten to one hundred times higher than those in nearby Neotropical continental land masses (Lugo et al. 1981). Both study sites have been subjected to intense human use in the past (Scatena 1989; García Montiel and Scatena 1994; Lugo et al. 1996; Molina Colón 1998). Uses ranged from selective cutting to outright conversion to agricultural and urban use. A third set of questions would be: How do these ecosystems recover from human disturbance? To what extent do they return to original conditions in terms of structure, functioning, and species composition?

Many changes to site conditions are introduced when forested lands are converted to other land uses such as agriculture, pastures and housing, or production used for charcoal. For example, aboveground biomass and soil organic matter are removed, topsoil is lost, remaining soil is compacted, the microenvironment is radically altered, and alien species can be introduced. As a result, forest regeneration can be arrested or the direction of succession can be altered (Ewel 1980). These effects can be chronic for as long as the nonforest land use continues. When the site is allowed to return to forest, the effect of land use change on succession can be analogous to that of a natural LID.

When a particular land use is abandoned, tree regeneration is mostly by seed from seed rain from adjacent sites (Aide et al. 1996; Molina Colón 1998). If there are legacies in the form of scattered trees or organic-matter-rich soils with residual root systems, these legacies expand their range by vegetative or sexual reproduction. On open and degraded areas, there is strong competition between invading life forms such as grasses, shrubs, or trees, and there is ample opportunity for invasion by alien species. Recovery processes are usually slow and regeneration leads to altered species composition relative to what was originally present at the site. The succession of both dry and wet forests has been shown to be slower on degraded sites than after clearcuts on sites of productive forests that are quickly allowed to recover (Aide et al. 1995; Molina Colón 1998).

The following trends emerge from the speed of recovery (resilience) of the dry forest after one hundred years under a variety of land uses (table 8.6):

1. Stands used for charcoal production (charcoal pits), which only had branches and stems removed without other stand modification, exceeded the mature forest in the majority of parameters considered and were over 66 percent recovered in all others after forty-five years of recovery. The same result was observed in wet forest previously subjected to charcoal extraction (García Montiel and Scatena 1994).
2. The baseball park, where the forest was removed and its soil compacted by human activity, had the slowest recovery but exceeded the mature forest in flower and fruit fall.
3. Regardless of previous land use, biomass and basal area were the parameters with the slowest level of recovery after forty-five years of growth.
4. In general, forest recovery forty-five years after human intervention was fairly advanced on most land uses; however, the rate of recovery was slow rel-

Table 8.6. Recovery of structure and functional processes in forty-five-year-old stands of dry forest that had been used for over one hundred years in the stated land use. Results are expressed in percentage of values measured in a nearby unconverted mature forest

	Past Land Use			
Parameter	Baseball Park	Charcoal Pits	Farmlands	Houses
STRUCTURE				
Basal area	42	111	68	53
Tree density	70	114	66	58
Tree height	59	85	76	76
Crown cover	58	75	75	67
Aboveground biomass	16	99	34	29
Root biomass	60	89	51	39
Species richness	85	128	115	90
Species evenness	80	100	92	60
Species heterogeneity	75	110	90	50
PROCESS				
Leaf fall	75	87	79	90
Woodfall	92	99	196	189
Reproductive part fall	114	66	154	211
Total litterfall	80	87	100	113

Source: Molina Colón 1998.

Note: Data are for trees with dbh greater than 2.5 centimeters.

ative to observations by Murphy et al. (1995) in experimental clearcuts where the stand was not chronically impacted as was the case with the land uses in table 8.6. Murphy et al. (1995) estimated that fifty years were required to reach maturity in most ecosystem parameters. Stands in table 8.6 are forty-five years old and not close to achieving maturity.

Wet forest stand succession on abandoned pastures after forty years of growth had recovered structural parameters (density, basal area, species diversity) to the point of being very similar to natural stands, but the speed of recovery was slower than when stands were recovering from only mechanical damage (Aide et al. 1995, 1996). In another study, Taylor et al. (1995) followed the recovery of a wet forest after three months of exposure to ionizing irradiation with a Cs^{137} source. After twenty-three years of growth, the biomass of saplings was less than the initial value. Species composition also failed to recover to initial conditions in wet forests recovering from anthropogenic disturbances (Taylor et al. 1995; Aide et al. 1996). Molina Colón (1998) made the same observation in the dry forest.

Species Invasions

The types of species that invade degraded sites, in other words, alien or invasive species, and the probability that these species might prevail over the long term, can be a biotic disturbance. These events lead to a fourth set of resilience questions. Do changes in species composition translate to changes in ecosystem functioning or resilience? Is it possible that when species composition changes either resilience or ecosystem functioning change but not both? To what extent do species substitutions occur without change in the state or type of the ecosystem?

These questions can be approached from two points of view: that those species changes that alter the species dominance of forests by definition involve a change to another state where the ecosystem has equal, less, or more resilience; or that changes in species composition only make a difference if there is a shift to species of different life forms or functional attributes (Ewel and Bigelow 1996). In either case it will be necessary to show that ecosystem functioning and/or level of resilience has shifted as a result of the species change.

There is evidence of changes in species composition in forests as a result of disturbances, particularly anthropogenic disturbances. The best evidence is from invasion of alien species. Such invasions leave no doubt of the change in species composition as a result of the disturbance event. Scatena et al. (1996) observed small-scale invasions of alien species in wet forests affected by Hurricane Hugo. These species invasions occurred early in the recovery of the forest and included species that had occupied the site when it was used intensively by humans. Similar observations were made by Molina Colón (1998) in her analysis of species composition of dry forests previously used for houses and farms.

Scatena and Lugo (1995) could not detect any large-scale changes in species composition in the wet forest as a result of Hurricane Hugo. Neither could Taylor et al. (1995) after the wet forest was exposed to ionizing radiation. However, when sites had been used extensively by humans, for example for pastures, the invasion of alien species into wet forest stands was more pervasive and included trees (Aide et al. 1996). Molina Colón (1998) observed the same phenomena in the dry forest.

Aide et al. (1996) suggested that it may be centuries before the original species contingent reclaims the wet forest site. An alternative view is that the ecosystem has fundamentally changed in terms of species composition but not in terms of function or structure. If so, it is unlikely that the original group of species will recover on the site. Instead, forest resilience mechanisms maintain the forest functioning with a different contingent of species. This "new ecosystem" was established because the nature of the disturbance (in this case chronic anthropogenic modification of the site) so changed conditions, that other species could grow when some of the original contingent of species could not. The new ecosystem can have altered nutrient or organic matter pools. For example, the presence of species like *Inga vera* and *Musa* sp. in the wet forest is a long-term legacy from past land uses. Their establishment resulted in changes in the nitrogen and potassium (respectively) pools of the stands (Scatena et al. 1996).

Changes in species composition illustrate at least two things about ecosystem resilience: (1) LIDs can be agents of change in species composition through their change in environmental conditions and opening of sites to competition and opportunity for invasion. When the disturbance is not severe, adjustments of species importance values without changes in species composition (Fu et al. 1996) are sufficient to recover ecosystem processes. (2) If the disturbance is severe enough (i.e., it's a LID), species invasions occur, including the invasion of alien species. If the change in the environment is such that changes in species are sufficient to maintain functioning, the system retains all its other essential structural characteristics (Aide et al. 1996). However, if the change is severe, a change in life form may occur (for example, a change from trees to grasses) and a full ecosystem bifurcation may take place.

Synergy between Anthropogenic and Natural Disturbances

We know that the dry and wet forests have been exposed to both human and natural disturbances (Willig et al. 1996; Molina Colón 1998). The fifth set of questions is: How do these different types of disturbances interact and modify each other? How do ecosystems maintain resilience under the combined effects of human and natural disturbances? A related issue is whether or not forest structure and/or function are related to the various disturbances.

The previous discussion on the changes in species composition, invasion of aliens, and changes in life forms illustrates the synergy between anthropogenic and natural disturbances because it revealed how the legacy of human use of the land continues to be visible in spite of long periods of recovery with or without further natural disturbances. This issue was approached by Willig et al. (1996) in terms of the functional diversity of bacterial communities in wet forest sites with legacies of anthropogenic and hurricane disturbances. They suggested that contemporary studies of spatial heterogeneity must account for the historical patterns of anthropogenic and natural disturbances to avoid spurious or incorrect conclusions.

Ecosystem-Level Characteristics That Result in Resilience

So far we have outlined biotic responses of subtropical wet and subtropical dry forest ecosystems to four different types of disturbances and the combinations among them. The disturbances were wind, soil moisture, land use change, species invasions, and various combinations such as wind and excessive rainfall during hurricanes and after landslides or the combination of anthropogenic and natural events. In this section, we generalize ecosystem responses to LIDs of all kinds. We believe these parameters to be the key sources of ecosystem resilience regardless of location.

Belowground Nutrients

Larger accumulation of nutrients and carbon below ground than above ground in both forest types provides a safety margin against acute disturbances that remove aboveground mass and nutrients from these systems (table 8.7). This high proportion of belowground storage provides resilience because it serves as a reservoir to replenish nutrient losses due to excessive export, mechanical

Table 8.7. Percentage of the total nutrient or carbon pool that is below ground in two subtropical forest stands in the Caribbean. Fine litter is included as above ground

Nutrient	Dry Forest[a]	Wet Forest[b]
N	95	95
P	99	99
K	97	98
C	76–81	52[c]–69

Sources: Data are from Lugo and Murphy 1986, Murphy and Lugo 1986b, Odum 1970, and Lugo 1992. Silver et al. 1994 contain similar data for extractable nutrients in the wet forest.

[a] Fine roots and/or soil organic matter to 0.85 meter.

[b] Depth to 1 meter for nutrients, 0.25 to 0.85 meter for soil organic matter, and 0.6 meter for fine roots.

[c] Includes all root sizes.

destruction of aboveground structure, some types of land use conversion, high rates of leaching, or clearcuts. These large belowground stores are examples of slow state variables that stabilize overall ecosystem function when the system is perturbed by events that remove aboveground materials from the site (Gunderson et al.1995). High amounts of soil organic matter in these forests (table 8.2) has an important role in nutrient retention and replenishment after LIDs (Silver et al. 1996c).

Rapid Fluxes

Rapid fluxes of nutrients, mass, and populations add resilience to systems faced with frequent or continuous disturbance events. Rapid fluxes of nutrients and mass allow for fast recovery of lost biomass in damaged forest stands, while fast rates of growth allow for the recovery of depleted population impacted by a LID. This resilience mechanism was demonstrated by Scatena (1995), who calculated 160 turnover periods (the ratio of standing stocks to fluxes) and fifty-five recovery periods (time required to be obliterated or returned to the original form or rate) of biotic and abiotic reservoirs of the wet forest. He found that the turnover of biological reservoirs by nutrient cycling was faster than their turnover by disturbances; in other words, the recovery of forest structure following tree falls, landslides, and hurricanes was faster than forest turnover by these disturbances. This shows high resilience because it allows forest stands to replenish nutrients and mass in the time interval between disturbances.

The ratio of total biomass to aboveground net primary productivity, a measure of replacement time, is high in mature wet and dry forests. For example, from the biomass stocks in Odum (1970) and Murphy and Lugo (1986b), and net primary productivity values in table 8.2 one finds a ratio of thirteen and twenty-three years for mature dry and wet forests, respectively. Considering the range of biomass involved (90 to more than 200 megagrams per hectare), these replacement times are fast. The value is faster in the dry forest because it has a low biomass and its aboveground net primary productivity is moderate (table 8.2). This forest stand is still accruing biomass (Murphy and Lugo 1986b). However, after Hurricane Hugo, the heavily disturbed wet forest produced aboveground biomass at a rate of 21 megagrams per hectare per year (Scatena et al. 1996), sufficient to replenish all the mature forest biomass in eleven years. Similar results in terms of basal area, tree density, and height growth are given in Fu et al. (1996) for another wet forest stand in the Luquillo Mountains. These recovery rates of total biomass, whether estimated at maturity or after a disturbance, show that fast fluxes of biomass are a resilience mechanism in these subtropical forest stands.

Fast movement of nutrients and biomass exchange and short generation times in organisms are mechanisms that provide flexibility and speed of response to perturbations. This allows systems to quickly "bounce back" when perturbed. This resilience is dependent on the availability of resources and propagules to use those resources.

Measured recovery rates after experimental or documented natural or anthropogenic disturbances reinforce results from the estimates of flux rates. The resilience is equally steep in subtropical dry and subtropical wet forests (tables 8.4 to 8.6). The resources to sustain these bursts of growth are stored belowground (table 8.7). The availability of propagules to take advantage of the growth opportunity is a function of surviving legacies (either resprouting [Silver and Vogt 1993] or seed, seedling, or saplings banks [Lugo and Zimmerman in press]) or immigration from other levels in the hierarchy.

Biotic Controls

Biotic control of fluxes, together with the multiplicity of alternative biotic pathways for the flow of energy and matter, is a resilience mechanism against those disturbances that chronically erode the resource capital of the ecosystem. For example, leaching of nutrients is a constant threat to the nutrient capital in high rainfall areas (Lugo and Scatena 1995). Fungal populations with extensive hyphal networks and large temporary nutrient storage capacity (Lodge 1996), high rates of retranslocation that allow nutrient reuse without exposing nutrients to loss (table 8.4), epiphytic organisms that remove nutrients from rain or the atmosphere, large nutrient accumulation in tissue, soil organic matter, and tree unions are examples of biotic mechanisms that counter this trend. The result of these mechanisms is that nutrients remain under biotic control and the opportunities for their loss are reduced.

Nutrient use-efficiency, which measures the production of biomass per unit nutrient used, is high for P and low for N in both forest types (table 8.4). When nutrients become scarce, use-efficiency increases. This behavior provides resilience because it allows within-stand nutrient reuse, increases the amount of biomass produced per unit nutrient invested, and minimizes nutrient losses. This also allows a steadier rate of organic matter processing while reducing potential losses of nutrients. However, when nutrients become available in large quantities, as they do after a hurricane disturbance, wet forests are capable of high immobilization rates at low nutrient use-efficiency (Scatena et al. 1996).

The shift in use-efficiency with a shift in nutrient availability underscores the importance of biotic control to ecosystem function and resilience. In the dry forest, crassulacean acid metabolism of succulent plants reduces the opportunity of water loss during photosynthesis, and the high water use-efficiency of plants assures optimal yield from a critically limiting factor (Lugo et al. 1978).

Species Richness and Redundancy

High species richness at ecosystem interfaces (redundancy) has been proposed by Silver et al. (1996 a,b) as important to resilience. They hypothesized that species redundancy at the atmospheric interface contributes to resilience because any species that survives and expands after a disturbance can recover the water and nutrient uptake function, even before the full complement of epiphytic species recovers at the interface. The interface with the atmosphere in the wet forest is occupied by a high species richness of epiphytic microfungi, lichens, algae, and plants that remove nutrients from the atmosphere (Lugo and Scatena 1995). The subtropical dry forest had a low richness of epiphyte species, but a high biomass of the dominant *Tillandsia recurvata*. These epiphytes also scavenge nutrients from the atmosphere. These scavenged nutrients eventually become available to trees with adventitious roots or through recycling from leachates or litterfall. Clusters of high species richness or high metabolic activity by a few species also occur in riparian zones at the interface between the terrestrial and aquatic systems.

Negative Feedbacks

Negative feedbacks occur at all levels of biotic interaction. Negative feedbacks provide stabilization to ecosystem processes (von Bertalanffy 1968) and help maintain homeostasis under changing environmental conditions. The storage function for matter and nutrients in litter, soil, wood, and the like all have built-in negative feedbacks (Odum 1983). These variables stabilize the ecosystem by dampening oscillations introduced by LIDs. Food chains in the subtropical wet forest contain numerous loops and negative feedbacks that also contribute to their resilience. Reagan et al. (1996) pointed out that these feeding loops among vertebrates and invertebrates were consistent with feedback mechanisms that control and stabilize populations in forests exposed to LIDs.

High Species Turnover and Self-Design

After mechanical disturbance, both forest types exhibited a rapid rate of species turnover and the same temporal pattern of species change at the less-than-one-hectare scale (figure 8.3). Following land use change, both forests were invaded by alien species and both maintained familiar physiognomy and functioning with the new species combinations. Even in disturbed sites where alien species

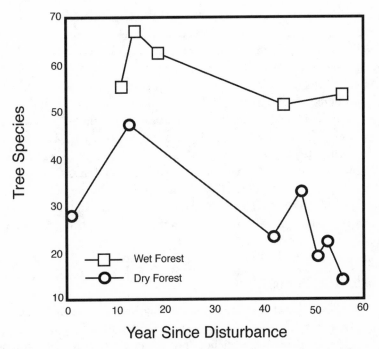

Figure 8.3. Patterns of species change after disturbances in two subtropical forests in Puerto Rico. The wet forest data were for a single 0.4-hectare plot responding to a hurricane event in 1932. The dry forest data represent a variety of disturbance events (experimental clearcuts and anthropogenic events based on different land uses) and different study plots with a range in area from 0.1 to 0.2 hectare. The minimum stem diameter varied among studies. The figure is illustrative of only the patterns of change.

attained high importance values, the forest condition was retained. It appeared as if, through natural self-design, these forests develop familiar levels of structure and velocity of processes with different combinations of species and different proportions of species importance values. The ability to remain within the pre-disturbance range of structures and functioning in spite of severe disturbances suggests that the changes in species composition and species turnover are a resilience mechanism. Alternatively, the system is controlled by physical variables such as the storage function described earlier, and species composition and turnover are of secondary importance.

Changes of ecosystem types, for instance, from forest to herbaceous, occur in large-scale landslides in wet forests and in highly degraded soils in dry forests. These changes are associated with loss of soil structure and fertility, or in other words, loss of organic matter and surface horizons. Species life forms change from trees to grasses, sedges, and/or ferns depending on moisture conditions. Dry areas only display grasses, while wet areas can have all three life forms. These changes in stability domains can be permanent or temporary depending on the permanence of the altered edaphic conditions.

We concluded with a listing of six characteristics of tropical forest ecosystems that contribute to their resilience. We believe that these are common to a wide range of tropical forests, regardless of their species composition, because they all share a similar function, which is the essence of an ecosystem. This was made clear by Evans (1956, p. 1127), who wrote, "an ecosystem involves the circulation, transformation, and accumulation of energy and matter through the medium of living things and their activities." The six attributes of resilience that we identify address the circulation, transformation, and accumulation of energy and matter of forest ecosystems rather than the medium of living things, which we had addressed earlier.

Concluding Remarks

Our objective in this chapter was to use comparisons to develop a framework for uncovering similarities of ecosystem function regardless of geographic location. We compared the resilience of dry forests with that of wet forests in Puerto Rico. The forests differed in substrate (limestone versus volcanic), annual rainfall (860 versus 3537 millimeters), and elevation (125–145 meters versus 150–500 meters), but they had similar hurricane and anthropogenic disturbance regimes, latitudinal belts, and biotemperatures. Overall stand structure and functioning and responses to water availability differed substantially, but eight stand attributes were similar in both forests: smooth canopy without emergent trees, high dominance of tree species, total number of tree species, number of tree species per unit area, annual rates of leaf fall, stemwood volume, and biomass production, and tree density for stems greater than 4 centimeters dbh. Although the ratio of total biomass to aboveground net primary productivity at maturity was lower in the dry forest (thirteen years) than in the wet forest (twenty-three years), the time required to restore full stand biomass after a clearcut was similar in both forests, as was their susceptibility to plant species invasions after anthropogenic land use changes. Timeline response curves of species richness, tree density, basal area, and leaf area index had the same patterns after clearcuts, but had different rates and end points.

The two forest types had similar levels of resilience. In addition to the cumulative effects of the responses of individuals to their environments (e.g., resprouting ability, short time to first reproduction), we propose the following parameters as the sources of forest resilience: high belowground storage of organic matter and nutrients; rapid fluxes of biomass, nutrients, and populations; biotic control of fluxes; high species richness and/or high metabolic activity at interfaces where external disturbance forces interact with the biota; negative feedbacks; high species turnover at a scale of less than 1 hectare; and the capacity of plant populations to form communities with similar structure and function despite different species composition (a property we call self-organization or self-design).

Acknowledgments

This work was done in cooperation with the University of Puerto Rico. It is part of the USDA Forest Service contribution to the National Science Foundation Long-Term Ecological Research Program at the Luquillo Experimental Forest (Grant BSR-8811902) to the Institute for Tropical Ecosystem Studies of the University of Puerto Rico, and the International Institute of Tropical Forestry, USDA Forest Service). We thank S. Brown, D. J. Lodge, E. Medina, A. Murphy, I. Ramjohn, and S. van Bloem for the review of the manuscript. Mildred Alayón and I. Ruiz contributed to the production of the manuscript.

Literature Cited

Aide, T. M., J. K. Zimmerman, L. Herrera, M. Rosario, and M. Serrano. 1995. Forest recovery in abandoned tropical pastures in Puerto Rico. *Forest Ecology and Management* 77:77–86.

Aide, T. M., J. K. Zimmerman, M. Rosario, and H. Marcano. 1996. Forest recovery in abandoned cattle pastures along an elevational gradient in northeastern Puerto Rico. *Biotropica* 28:537–548.

Basnet, K., F. N. Scatena, G. E. Likens, and A. E. Lugo. 1993. Ecological consequences of root grafting in tabonuco (*Dacryodes excelsa*) trees in the Luquillo Experimental Forest, Puerto Rico. *Biotropica* 25:28–35.

Benedict, F. F. 1976. Herbivory rates and leaf properties in four forests in Puerto Rico and Florida. Master's thesis. Department of Botany, University of Florida, Gainesville.

Brokaw, N. V. L., and J. S. Grear. 1991. Forest structure before and after Hurricane Hugo in Luquillo Experimental Forest, Puerto Rico. *Biotropica* 23:386–392.

Chinea, J. D. 1980. The forest vegetation of the limestone hills of northern Puerto Rico. Master's thesis. Cornell University, Ithaca, N.Y.

Crow, T. R. 1980. A rainforest chronicle: A thirty-year record of change in structure and composition at El Verde, Puerto Rico. *Biotropica* 12:42–55.

Domínguez Cristóbal, C. 1989a. Situación forestal de Puerto Rico durante el siglo XIX. *Acta Científica* 3:24–25.

Domínguez Cristóbal, C. M. 1989b. Situación forestal pre-hispánica de Puerto Rico. *Acta Científica* 3:63–66.

————. 1989c. Situación forestal de Puerto Rico durante el siglo 16. *Acta Científica* 3:67–70.

————. 1989d. Situación forestal de Puerto Rico durante el siglo 17. *Acta Científica* 3:71–72.

————. 1989e. Situación forestal de Puerto Rico durante el siglo 18. *Acta Científica* 3:73–76.

————. 1989f. Situación forestal de Puerto Rico durante el siglo 20 (hasta 1975). *Acta Científica* 3:77–82.

Evans, F. C. 1956. Ecosystem as the basic unit in ecology. *Science* 123:1127–1128.

Ewel, J. J. 1977. Differences between wet and dry successional tropical ecosystems. *Geo-Eco-Trop* 1:103–117.

————. 1980. Tropical succession: Manifold routes to maturity. *Biotropica* 12(supplement):2–7.

Ewel, J. J., and S. W. Bigelow. 1996. Plant life forms and tropical ecosystem functioning. Pp. 101–126 in *Biodiversity and ecosystem processes in tropical forests*, edited by G. Orians, R. Dirzo, and J. H. Cushman. Berlin: Springer-Verlag.

Fu, S., C. Rodríguez Pedraza, and A. E. Lugo. 1996. A twelve-year comparison of stand changes in a mahogany plantation and a paired natural forest of similar age. *Biotropica* 28:515–524.

García-Martinó, A. R., G. S. Warner, F. N. Scatena, and D. L. Civco. 1996. Rainfall, runoff and elevation relationships in the Luquillo Mountains of Puerto Rico. *Caribbean Journal of Science* 32:413–424.

García Montiel, D., and F. N. Scatena. 1994. The effect of human activity on the structure and composition of a tropical forest in Puerto Rico. *Forest Ecology and Management* 63:57–78.

Gunderson, L. H., C. S. Holling, and S. S. Light. 1995. Barriers broken and bridges built: A synthesis. Pp. 489–532 in *Barriers and bridges to renewal of ecosystems and institutions*, edited by L. H. Gunderson, C. S. Holling, and S. S. Light. New York: Columbia University Press.

Guzmán-Grajales, S. M., and L. R. Walker. 1991. Differential seedling responses to litter after Hurricane Hugo in the Luquillo Experimental Forest of Puerto Rico. *Biotropica* 23:407–413.

Holdridge, L. R. 1967. *Life zone ecology*. San José, Costa Rica: Tropical Science Center.

Jordan, C. F. 1971. Production of a tropical forest and its relation to world pattern of energy storage. *Journal of Ecology* 59:127–142.

Kepler, C. B., and A. K. Kepler. 1970. Preliminary comparison of bird species diversity and density in Luquillo and Guánica forests. Pp. E183–E191 in *A tropical rain forest*, edited by H. T. Odum and R. F. Pigeon. Springfield, Va.: National Technical Information Service.

Klinge, H., W. A. Rodríguez, E. Bruning, and E. J. Fittkau. 1975. Biomass and structure in a Central Amazonian rain forest. Pp. 115–122 in *Tropical ecological systems*, edited by F. B. Golley and E. Medina. New York: Springer-Verlag.

Little, E. L., Jr., R. O. Woodbury, and F. H. Wadsworth. 1974. *Trees of Puerto Rico and the Virgin Islands*. USDA Forest Service Agriculture Handbook no. 449. Washington, D.C.

Lodge, D. J. 1996. Microorganisms. Pp. 53–108 in *The food web of a tropical rain forest*, edited by D. P. Reagan and R. B. Waide. Chicago: University of Chicago Press.

Lodge, D. J., F. N. Scatena, C. E. Asbury, and M. J. Sánchez. 1991. Fine litterfall and related nutrient inputs resulting from Hurricane Hugo in subtropical wet and lower montane rain forests of Puerto Rico. *Biotropica* 23:336–342.

Lugo, A. E. 1978. Stress and ecosystems. Pp. 62–101 in *Energy and environmental stress in aquatic ecosystems*, edited by J. H. Thorp and J. W. Gibbons. U.S. Department of Energy Symposium Series (CONF 77114). National Technical Information Services, Springfield, Va.

———. 1988. Ecological aspects of catastrophes in Caribbean islands. *Acta Científica* 2:24–31.

———. 1991. Dominancia y diversidad de plantas en Isla de Mona. *Acta Científica* 5:65–71.

———. 1992. Comparison of tropical tree plantations with secondary forests of similar age. *Ecological Monographs* 62:1–41.

———. 1996. Caribbean island landscapes: Indicators of the effects of economic growth on the region. *Environment and Development Economics* 1:128–136.

Lugo, A. E., J. Figueroa Colón, and F. N. Scatena. 1999. The Caribbean. Pp. 593–622 in *North American terrestrial vegetation*. 2nd ed. Edited by M. G. Barbour and D. D. Billings. Cambridge: Cambridge University Press.

Lugo, A. E., J. A. González Liboy, B. Cintrón, and K. Dugger. 1978. Structure, productivity, and transpiration of a subtropical dry forest in Puerto Rico. *Biotropica* 10:278–291.

Lugo, A. E., and P. G. Murphy. 1986. Nutrient dynamics of a Puerto Rican subtropical dry forest. *Journal of Tropical Ecology* 2:55–76.

Lugo, A. E., O. Ramos, S. Molina, F. N. Scatena, and L. L. Vélez Rodríguez. 1996. *A fifty-three year record of land use change in the Guánica forest biosphere reserve and its vicinity*. International Institute of Tropical Forestry, USDA Forest Service, Río Piedras, Puerto Rico.

Lugo, A. E., and F. N. Scatena. 1992. Epiphytes and climate change research in the Caribbean: A proposal. *Selbyana* 13:123–130.

———. 1995. Ecosystem-level properties of the Luquillo Experimental Forest with emphasis on the tabonuco forest. Pp. 59–108 in *Tropical forests: Management and ecology*, edited by A. E. Lugo and C. Lowe. New York: Springer-Verlag.

———. 1996. Background and catastrophic tree mortality in tropical moist, wet, and rain forests. *Biotropica* 28:585–599.

Lugo, A. E., R. Schmidt, and S. Brown. 1981. Tropical forests in the Caribbean. *Ambio* 10:318–324.

Lugo, A. E., and J. K. Zimmerman. In press. Ecological life histories of tropical trees with emphasis on disturbance effects. *Tropical tree seed manual*, edited by J. Vozzo. USDA Forest Service, Southern Research Station, Asheville, N.C.

McDowell, W. H., C. P. McSwiney, and W. B. Bowden. 1996. Effects of hurricane disturbance on groundwater chemistry and riparian function in a tropical rain forest. *Biotropica* 28:577–584.

Molina Colón, S. 1998. Long-term recovery of a Caribbean dry forest after abandonment of different land uses in Guánica, Puerto Rico. Ph.D. diss., University of Puerto Rico, Río Piedras.

Murphy, P. G., and A. E. Lugo. 1986a. Ecology of tropical dry forest. *Annual Review of Ecology and Systematics* 17:67–88.

————. 1986b. Structure and biomass of a subtropical dry forest in Puerto Rico. *Biotropica* 18:89–96.

Murphy, P. G., A. E. Lugo, A. J. Murphy, and D. C. Nepstad. 1995. The dry forests of Puerto Rico's south coast. Pp. 178–209 in *Tropical forests: Management and ecology*, edited by A. E. Lugo and C. Lowe. New York: Springer-Verlag.

Neumann, C. J., G. W. Caso, and B. R. Jaruinen. 1978. Tropical cyclones of the North Atlantic Ocean, 1871–1978. U.S. Department of Commerce, National Oceanic and Atmospheric Administration, National Climatic Center. Asheville, N.C.

Odum, H. T. 1970. Summary: An emerging view of the ecological system at El Verde. Pp. I191–I281 in *A tropical rain forest*, edited by H. T. Odum and R. F. Pigeon. National Technical Information Service, Springfield, Va.

————. 1983. *Systems ecology: An introduction.* New York: John Wiley and Sons.

Odum, H. T., G. Drewry, and J. R. Kline. 1970. Climate at El Verde. 1963–1966. Pp. 347–418 in *A tropical rain forest*, edited by H. T. Odum and R. F. Pigeon. National Technical Information Service, Springfield, Va.

Parrotta, J. A., and D. J. Lodge. 1991. Fine root dynamics in a subtropical wet forest following Hurricane Hugo disturbance in Puerto Rico. *Biotropica* 23:343–347.

Quigley, M. F. 1994. Latitudinal gradients in temperate and tropical seasonal forests. Ph.D. diss. Louisiana State University, Baton Rouge.

Reagan, D. P., G. R. Camilo, and R. B. Waide. 1996. The community food web: Major properties and patterns of organization. Pp. 461–488 in *The food web of a tropical rain forest*, edited by D. P. Reagan and R. B. Waide. Chicago: University of Chicago Press.

Scatena, F. N. 1989. An introduction to the physiography and history of the Bisley Experimental Watersheds in the Luquillo Mountains of Puerto Rico. General Technical Report SO-72. USDA Forest Service Southern Forest Experiment Station, New Orleans, La.

————. 1995. Relative scales of time and effectiveness of watershed processes in a tropical montane rain forest of Puerto Rico. Pp. 103–111 in *Natural and anthropogenic influences in fluvial geomorphology*. Geophysical Monograph 89. Washington, D.C.: American Geophysical Union.

Scatena, F. N., and M. C. Larsen. 1991. Physical aspects of Hurricane Hugo in Puerto Rico. *Biotropica* 23:317–323.

Scatena, F. N., and A. E. Lugo. 1995. Geomorphology, disturbance, and the soil and vegetation of two subtropical wet steepland watersheds of Puerto Rico. *Geomorphology* 13:199–213.

Scatena, F. N., S. Moya, C. Estrada, and J. D. Chinea. 1996. The first five years in the reorganization of aboveground biomass and nutrient use following Hurricane Hugo in the Bisley Experimental Watersheds, Luquillo Experimental Forest, Puerto Rico. *Biotropica* 28:424–440.

Silver, W., F. N. Scatena, A. H. Johnson, T. G. Siccama, and F. Watt. 1996c. At what temporal scales does disturbance affect background nutrient pools? *Biotropica* 28:441–457.

Silver, W. L. 1994. Is nutrient availability related to plant nutrient use in humid tropical forests? *Oecologia* 98:336–343.

Silver, W. L., S. Brown, and A. E. Lugo. 1996a. Effects of changes in biodiversity on ecosystem function in tropical forests. *Conservation Biology* 10:17–24.

————. 1996b. Biodiversity and biogeochemical cycles. Pp. 49–67 in *Biodiversity and ecosystem process in tropical forests*, edited by G. Orians, R. Dirzo, and J. H. Cushman. Heidelberg: Springer-Verlag.

Silver, W. L., A. E. Lugo, and M. Keller. 1999. Soil oxygen availability and biogeochemistry along rainfall and topographic gradients in upland wet tropical forest soils. *Biogeochemistry* 44:301–328.

Silver, W. L., F. N. Scatena, A. H. Johnson, T .G. Siccama, and M. J. Sanchez. 1994. Nutrient availability in a montane wet tropical forest: Spatial patterns and methodological considerations. *Plant and Soil* 164:129–145.

Silver, W. L., and K. A. Vogt. 1993. Fine root dynamics following single and multiple disturbances in a subtropical wet forest ecosystem. *Journal of Ecology* 81:729–738.

Smith, R. F. 1970. The vegetation structure of a Puerto Rican rain forest before and after short-term gamma irradiation. Pp. D103–D140 in *A tropical rain forest*, edited by H. T. Odum and R. F. Pigeon. National Technical Information Service, Springfield, Va.

Taylor, C. M., S. Silander, R. B. Waide, and W. J. Pfeifer. 1995. Recovery of a tropical forest after gamma irradiation: A twenty-three-year chronicle. Pp. 258–285 in *Tropical forests: Management and ecology*, edited by A. E. Lugo and C. Lowe. New York: Springer-Verlag.

Thomlinson, J. R., M. I. Serrano, T. del M. López, T. M. Aide, and J. K. Zimmerman. 1996. Land-use dynamics in a post-agricultural Puerto Rican landscape (1936–1988). *Biotropica* 28:525–536.

von Bertalanffy, L. 1968. *General system theory*. New York: George Braziller.

Walker, L. R. 1995. Timing of post-hurricane tree mortality in Puerto Rico. *Journal of Tropical Ecology* 11:315–320.

Walker, L. R., N. V. L. Brokaw, and R. B. Waide, eds. 1991. Ecosystem, plant, and animal responses to hurricanes in the Caribbean. *Biotropica* 23:313–521.

Weaver, P. L. 1987. Structure and dynamics in the Colorado forests of the Luquillo Mountains of Puerto Rico. Ph.D. diss. Michigan State University, East Lansing.

Weaver, P. L. and P. G. Murphy. 1990. Forest structure and productivity in Puerto Rico's Luquillo Mountains. *Biotropica* 22:69–82.

Whigham, D. F., I. Olmsted, E. Cabrera Cano, and M. E. Harmon. 1991. The impact of Hurricane Gilbert on trees, litterfall, and woody debris in a dry tropical forest in northeastern Yucatan Peninsula. *Biotropica* 23:434–441.

Whittaker, R. H. 1965. Dominance and diversity in land plant communities. *Science* 147:250–260.

Wiegert, R. G. 1970. Effects of ionizing radiation on leaf fall, decomposition, and litter microarthropods of a montane rain forest. Pp. H89–100 in *A tropical rain forest*, edited by H. T. Odum and R. F. Pigeon. National Technical Information Service, Springfield Va.

Williams, G. C. 1966. *Adaptation and natural selection*. Princeton, N.J.: Princeton University Press.

Willig, M. R., D. L. Moorhead, S. B. Cox, and J. C. Zak. 1996. Functional diversity of soil bacterial communities in the tabonuco forest: Interaction of anthropogenic and natural disturbance. *Biotropica* 28:471–483.

Wunderle, J. M., D. J. Lodge, and R. B. Waide. 1992. Short-term effects of Hurricane Gilbert on terrestrial bird populations on Jamaica. *Auk* 109:148–166.

9

Forest Dynamics in the Southeastern United States: Managing Multiple Stable States

Garry D. Peterson

Land use conversion, urban development, and manipulation of disturbance regimes have dramatically altered the composition, abundance, and distribution of forests in the southeastern coastal plain of the United States. These practices have altered the resilience of historical landscapes, causing an open forest that frequently burns to be replaced by closed mesic forest that seldom burns. These changes have led land managers at Eglin Air Force Base in Florida to attempt to restore the historical landscape. The current landscape of the base is composed of a mosaic of open forest and mesic forest. In isolation each of these forest types will persist, but in combination their complex dynamics complicate the management of the landscape. Understanding ecological resilience (Holling 1996; Gunderson et al. chapter 1) is critical for management objectives such as restoration and preservation of the historical landscape. An analysis of the landscape dynamics of Eglin Air Force Base illustrates how the existence of alternate stable states alters the management, preservation, and restoration of ecosystems.

The chapter is organized in five sections. The first section introduces the study site, its forest types, their relationships in terms of species composition, disturbance regimes and succession, the history of the site, and a simple conceptual model of forest dynamics. The second section introduces a spatial simulation model. In the third section, the model is used to evaluate a set of management options. The fourth section discusses the results of these policy evaluations, and the final section suggests general implications for the management of landscapes that contain alternate stable states.

Pine-Oak Forest Dynamics in Northwest Florida

The forests of the southeastern United States are typically found on sandy soils characterized by relatively low nutrient concentrations and little organic matter, covering a range of hydrologic conditions from xeric to hydric, and subject to periodic disturbances of fire, storms, and pests. Prior to land conversion, a large percentage of the forests were dominated by longleaf pine (*Pinus palustris*). The longleaf pine forests can be considered a stable state, since frequent fires maintained the nearly monotypic dominance of longleaf. Other forest types are found in the landscape, including sand pine (*Pinus clausa*) and oak forests codominated by various species of *Quercus*. Each of these forests is briefly described.

Longleaf Pine

Formerly, forests of longleaf pine (*Pinus palustris*) covered much of the coastal plain in the southeastern United States. During the past two centuries, human activities such as logging, agriculture, and fire suppression have reduced the area covered by longleaf pine forest to less than 5 percent of its former range. The remnant patches of longleaf pine forest contain several species of endangered plants and animals, the most prominent of which is the red-cockaded woodpecker (*Picoides borealis*) (Jackson Guard 1993). Forests of oak and sand pine have replaced large areas of longleaf pine. The largest remaining area of contiguous old-growth longleaf pine forest is located in the northwest Florida, comprising 1,500 square kilometers of the roughly 1,900 square kilometers of Eglin Air Force Base (figure 9.1).

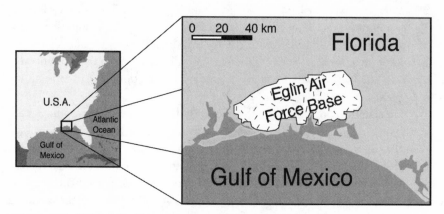

Figure 9.1. Eglin Air Force Base covers 1,900 square kilometers in northwestern Florida and is in the southeastern coastal plain physiographic region.

Mature stands of longleaf pine often produce open savanna ecosystems that possess an understory rich in grasses and herbs. The productivity and spatial distribution of the understory is determined by edaphic conditions of soil nutrients and water holding capacity. The understory was composed of grasses such as wiregrass (*Aristida* spp.) or palms such as saw palmetto (*Serenoa repens*). This savanna state was maintained by frequent fires that consumed the understory but rarely damaged the overstory trees. Indeed, the adaptations of longleaf pine, such as its thick bark and its reproductive strategy suggest evolution in a relatively frequent (less than fifteen years) return interval of fires.

Oaks, Pines, and Fires

The sandy, well-drained soils of northern Florida support forests dominated by either longleaf pine or various oak (*Quercus*) species (Abrahamson and Hartnett 1990). The frequency of ground fires usually determines which vegetation type dominates at a specific location (Heyward 1939; Rebertus et al. 1989).

Ground fires in the forest of north Florida consume fine fuels and kill the aboveground portion of understory and midstory vegetation, but they usually do not burn the canopy trees of the forest. When fire kills only the aboveground portion of the understory vegetation, the vegetation usually regenerates quickly. Within a year, or less in the case of spring fires, understory vegetation can regrow to such an extent that the accumulated fuel can once again propagate fire. After a site has gone several years without fire, its combustibility begins to decrease (Platt et al. 1988; Robbins and Myers 1992).

Fire plays a key role in determining what types of tree seedlings survive to become trees. Longleaf pines have adapted to regimes of infrequent ground fires. Young pines remaining in a "grass" stage for up to seven years, accumulating root biomass. Following this stage, pine seedlings grow into saplings within one year, in essence elevating vulnerable meristem tissue above ground-fire kill zones. Young oaks, however, are killed by fire. Hence, periods of no fire allow for the survival and growth of young oaks. Mature oaks form dense stands that reduce the amount of light that reaches the ground, producing a sparse understory. Oaks shed leaves that form a compressed, low-oxygen litter layer that suppresses the accumulation of understory vegetation and other potentially combustible detritus.

Fire frequency and intensity creates alternative stable states in these forests—one dominated by longleaf pine and the other by oak trees (Rebertus et al. 1989). The bark of longleaf pine is resistant to frequent, low intensity fire (temperature and duration). Fine fuels, comprising shed longleaf needles and resprouting understory, support more-frequent, less-intense fires. This fire regime reinforces the persistence of a longleaf pine savanna. Oak dominance inhibits fire, and the resulting absence of fire reinforces the oak-dominated state.

Sand Pine

A third tree species, sand pine (*Pinus clausa*), further complicates ecological dynamics in low-lying areas of northwest Florida. Sand pine historically occurred along the coast but in past decades has spread into areas formerly occupied by longleaf pine (Provencher et al. 1998). Sand pine in northwest Florida is the Choctawhatchee variety, which is ecologically distinct from other sand pine varieties. Although interior Florida sand pine are serotinous, requiring fires to open their cones before seeds can regenerate, the Choctawhatchee variety is not (Parker and Hamrick 1996). Sand pine is not as fire tolerant but reaches sexual maturity much faster than longleaf pine does. These differences mean that sites that are not regularly burned can be invaded by neighboring sand pine within several years. Sites dominated by sand pine are less combustible than sites composed primarily of hardwood species (Provencher et al. 1998). For the purposes of the model developed later in this chapter, I group sand pine–dominated sites with hardwood-dominated sites, because both forest types inhibit fire.

The interactions of longleaf pine, sand pine, and oaks can be captured in a simple model (figure 9.2). Northern Florida upland forest can be conceptualized as being composed of a mix of longleaf pine, hardwoods, and sand pine. Changes in the forest can be thought of as moving through a triangular state space whose axes are composed of the proportion of longleaf pine, the proportion hardwoods, and the proportion sand pine, and which in combination equal the total forest cover (i.e., longleaf % + hardwood % + sand pine % = 100%). On wet soils in low-lying areas, forest can be composed of either hardwoods or a mix of hardwoods and sand pine—the low topography prevents conversion to upland forest. Although most fires are ground fires, canopy fires can kill all or almost all trees within an area. Following such fires, a forest site may be temporarily converted to an area whose vegetation is composed of herbs, grasses, and young hardwoods. Longleaf pine–dominated sites burn easily, while hardwood and sand pine–dominated sites suppress fire. This interaction between fire and vegetation type means that the forest can be divided into two types: longleaf pine–dominated forest that burns frequently, and hardwood (or sand pine) dominated forest that burns infrequently. Analyzing the forest in terms of these two distinct relationships with fire usefully simplifies the analysis of landscape dynamics.

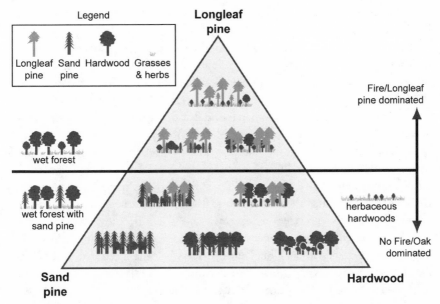

Figure 9.2. Forest dynamics can be simplified using a conceptual model of forest and fire dynamics. Forests can be dominated by longleaf pine, sand pine, or hardwoods as indicated by the vertices of the triangle, forests are often a mixture of these three as shown by the images inside the triangle. Forests on sandy soils are composed of a mix of longleaf pine, hardwoods, and sand pine. Forests on wet soils can be composed of either hardwoods or a mix of hardwoods and sand pine. Longleaf-dominated sites are maintained by frequent fires. Hardwood and sand pine sites burn less frequently.

Study Site—Eglin Air Force Base

Eglin Air Force Base covers an area of about 1,900 square kilometers in northwestern Florida (figure 9.1). The base contains about 1,500 square kilometers of longleaf pine forest, the largest remaining area of an ecosystem that formerly covered much of the southeastern United States. The land managers of Eglin Air Force Base wish to expand the area and increase the quality of the longleaf pine forest on the base in order to meet requirements of the federal Endangered Species Act and because of local concerns about fire and forest management (Jackson Guard 1993). Ecological management has been ongoing since the early 1990s. Burning the forest using prescribed fire has been one of the main tools of this activity. The use of prescribed fire is motivated by both ecological and safety concerns. Ecologically, prescribed burns are conducted to attempt to maintain and increase the area of longleaf pine forest while also reducing the risk of uncontrollable wildfire. However, the management of the forest is complicated by the interaction of fire and forest dynamics (Hardesty et al. 2000).

Restoring the forest requires understanding its ecology and its history. Ecological understanding is needed to develop useful management plans, while an understanding of the past is required to understand what processes and states should be encouraged or discouraged. Ecosystem management operates at the scale of thousands of square kilometers and decades. Eglin's land managers have a deep understanding of the land they manage; however, that understanding has necessarily been constrained to relatively small and short-term scales. Similarly, a large body of scientific literature on longleaf pine forest exists (Platt et al. 1988; Glitzenstein et al. 1995), but this knowledge has often been developed at the scale of forest stands over several years. Bridging this scale mismatch between the requirements of ecosystem management and scientific knowledge and management experience requires new methodologies. Through a partnership with The Nature Conservancy, Eglin's land managers became interested in developing simulation models of Eglin's forest dynamics to translate small-scale ecological understanding to management scales. After developing ecological models at management scales, these models were used to evaluate alternative management strategies.

Historically, longleaf pine forest dominated the site now occupied by Eglin Air Force Base. This landscape was maintained by wildfires, which produced large, homogeneous areas of longleaf savanna. Interspersed in the landscape, especially along stream networks, were areas of oak and sand pine, along with other areas where fire could not easily spread (figure 9.3a).

Today, the landscape is clearly anthropogenic. Logging, agricultural clearing, establishment of a military base, and the construction of roads and airstrips provided Eglin Air Force Base with a complex, diverse, and undocumented history of land use change. Fire suppression has reduced the number of fires, and landscape fragmentation has reduced the ability of the remaining fires to spread and burn large areas of the landscape. The reduction of fire frequency, especially in areas highly fragmented by stream networks and roads, has decreased the total area of longleaf pine. Some of the larger contiguous areas of longleaf pine were maintained, since they were ignited frequently enough to maintain the area in longleaf pine. However, most of the smaller areas of longleaf pine were invaded by hardwoods. Some of the wetter areas along the coast or near sand pine plantations were invaded by sand pine. The areas along roads were burned less frequently and so served as corridors for the invasion of both sand pine and hardwoods.

The current landscape configuration reduces the likelihood of large fires, because the spread of fire has been curtailed relative to the historical landscape.

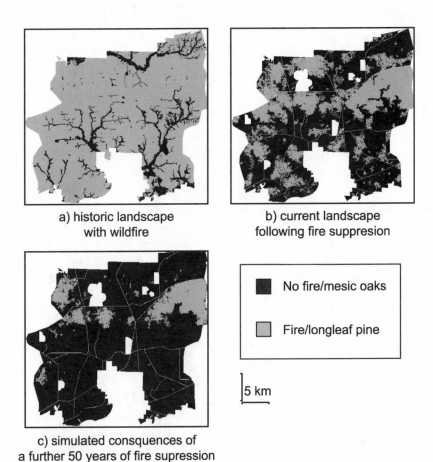

a) historic landscape
with wildfire

b) current landscape
following fire suppresion

No fire/mesic oaks

Fire/longleaf pine

5 km

c) simulated consquences of
a further 50 years of fire supression

Figure 9.3. Fire suppression has reduced the area of longleaf pine forest at the Eglin study site. (a) The historical landscape was primarily a longleaf pine savanna (data adapted from the Nature Conservancy). It was maintained by frequent wildfires. (b) Fire suppression and fragmentation converted a longleaf-dominated landscape to a landscape of mixed longleaf and oaks. (c) If historical management practices are continued, simulation models suggest further reduction of longleaf pine area.

Decades of fire suppression and fragmentation have led to a landscape that is composed of large areas of relatively noncombustible vegetation (figure 9.3b). Consequently, further fire suppression will continue to degrade the landscape. Fire will continue to divide the landscape into areas of longleaf pine- and oak- or sand pine–dominated areas (figure 9.3c). Longleaf pine will occur in patches that are large enough to have frequent lighting-initiated wildfires and homogeneous enough for fire to spread across a large portion of that patch. Areas open to fire along the edges of stream networks will be converted to longleaf, while areas surrounded by the stream network or roads, where it is difficult for fire to spread, will be converted to oak.

Modeling Landscape Dynamics

I used a spatial simulation model to examine the landscape dynamics of Eglin Air Force Base management under a variety of different strategies. This model represents the Eglin landscape as a 25 x 25-kilometer matrix of sites, each 60 meters on edge. This resolution was chosen for two reasons. First, it is a reasonable scale at which to develop management alternatives, displaying heterogeneity that is important to management while not providing overwhelming detail (Jackson Guard 1993). Second, tree size appears to have most of its variation at scales of less than 28 meters (Platt and Rathburn 1993), making 60 meters an ecologically reasonable scale to represent a portion of forest. Both the forest and the understory at each site were represented. Forest type was modeled as a set of discrete forest states, and the understory is represented by the time since the last two fires at a site. The time since fire implicitly represents combustibility of the understory vegetation. Vegetation change is controlled by a set of rules based upon succession, fire frequency, and seed dispersal.

The model operates at a yearly time scale. Each year fires occur and vegetation changes. Fires can be either prescribed fires, ignited by managers, or wildfires, ignited by lightning. The effectiveness of prescribed fire on a site within a burned area depends upon the vegetation type of that site and the time since that site was last burned. Wildfires are ignited by lightning. The ability of a fire to spread into a site depends on the vegetation type of that site and the time since that site was last burned. Remnant patches of unburned vegetation remain within a burned area, and fire spread is impeded by older, less combustible vegetation. The spread of fire can be halted by roads or difficult-to-burn vegetation.

This model was developed, tested, and refined through a series of workshops involving ecologists and land managers (Hardesty et al. 2000). At these workshops the model was used, discussed, tested, and revised. The model structure was based upon the literature on southern longleaf pine forests (Boyer and Peterson 1983; Rebertus et al. 1989; Platt and Rathburn 1993; Glitzenstein et al. 1995), ecologists' opinions on how the Eglin forest differed from other studied sites, and the experience of Eglin's land managers. The model was tested by assessing its ability to reproduce the historical landscape changes described above. Quantitative tests of the model were impossible, due to a lack of comparable data. The major tests of the model's credibility were its ability to reproduce and maintain the historical Eglin landscape, and its ability to reproduce the historical dynamics of hardwood and sand pine invasion into longleaf pine sites.

The model is described in detail elsewhere (Peterson 1999). For the analysis presented in this chapter, the key feature of the model is its representation of longleaf pine–dominated and hardwood-dominated forests as alternate stable states. The qualitative conclusions of this chapter are robust to changes in the model as long as longleaf pine–dominated and hardwood-dominated forests remain as alternative stable states.

Management Alternatives

Fire management requires a strategy for deciding when and where to burn. Using the model to evaluate land management strategies offers a way to test ideas about management—ideas that combine both social and ecological mental models—and to locate important uncertainities and unintended consquences of these strategies. Based upon discussions with land managers, three types of land management strategies were identified for the model. The first and second, rotation burning and "responsive" burning (described below), both use prescribed fire to manage the landscape, while the third strategy uses wildfire. These strategies capture the general form rather than the details of different management approaches. Before describing the three strategies in detail, I outline the management practice of prescribed fire.

Prescribed Fire

Land managers set prescribed fires. Weather, ignition pattern, the skill of the burn team, and other factors can all affect the results of any prescribed fire. Prescribed burns can be ignited both from the air and on the ground. Aerial ignitions, in which fire is dropped from helicopters, allow larger areas to be burned, but such fires vary more in intensity than do ground fires. Ground ignitions are started by using drip torches to establish a fire front that is allowed to burn over a plot and into a previously set *backing fire*—a fire that burns backward away from a fire break such as a road. Ground ignitions can provide more control over fire spread and can usually be managed to burn more evenly than aerial ignitions. The fire front of a ground fire provides more opportunities for a site's ignition than that of an aerial fire. Land managers estimate that when they burn a hardwood-invaded longleaf site, a ground burn will actually ignite about 75 percent of the area, while an aerial burn will ignite only about 60 percent of the area. However, it is difficult to start and control large fires from the ground.

Based on analysis of long-term prescribed fire data, Eglin managers were able to distinguish between the average size of fires started aerially and those started on the ground. Aerially ignited fires burned an average area of 24,000 acres (9 square kilometers), while the ground-ignited fires covered an average of 6,000 acres (2.25 square kilometers).

Prescribed burns are difficult to organize. They require trained fire managers and specific weather conditions. Eglin managers estimate that under current safety and military constraints, conditions are appropriate for burning only about eighty days per year, even though staff and funding levels would actually allow them to conduct a maximum of 120 burns per year, roughly evenly split between aerial and ground fires. These estimates provide an upper bound of sixty ground and sixty aerial prescribed fires per year over Eglin Air Force Base. Since the simulation area covers about one-third of the base, the modeled area could receive a maximum of twenty ground and twenty aerial burns per year, corresponding to a fire frequency of about 2.75 years per fire. Alternative fire management strategies were tested using this frequency of fire, and several slower rates (3.6, 5.5, and 11 years per fire).

Fire Management Strategies

The average fire frequencies for prescribed burns do not completely determine fire management strategies for a given landscape. The spatial application of prescribed fire is another crucial element. One common fire management strategy is simply to rotate prescribed fire across a landscape by dividing the landscape into designated *burn blocks* and burning each in turn, over some set period, with the same frequency.

An alternative method, developed during the workshop process specifically for Eglin Air Force Base (and therefore not a typical fire management strategy), is to burn specific burn blocks more frequently than others. This "responsive" strategy burns areas that contain relatively high proportions of longleaf pine more frequently than areas containing only small amounts of longleaf pine.

The simplest approach to fire management is to in fact do nothing, letting wildfires ignited by random lightning strikes burn without human intervention. This strategy is unlikely to be used at Eglin Air Force Base, but it was included as a baseline against which other fire management strategies could be compared. Two types of wildfire management were evaluated—a restoration of the historical rate of wildfire initiation and a suppressed rate of wildfire initiation.

These alternative management strategies were used to probe the dynamics of the managed Eglin landscape. Although such simulation exercises cannot be expected to predict the details of any specific management regime, they can identify general patterns that describe the dynamics of the forest communities of north Florida and help to pinpoint the gaps in our understanding of their ecology.

The success of these modeling strategies was evaluated based upon the changes they produced in the Eglin Air Force Base landscape over fifty years. The pattern of change produced by each strategy has a complex temporal dynamic; however, here I will only describe the coarse spatial changes. To simplify the analysis of changes in the spatial distribution of vegetation, a fixed time period of fifty years was used for analysis. The "future" landscape at the end of these fifty years was compared with the "present" landscape, and the changes were calculated. Fifty years was chosen, because it is long enough to detect the long-term consequences of a fire regime on the landscape but short enough to be relevant to management planning decisions. While management plans are often made at less than decadal time scales, forest management must evaluate these decade-length plans in terms of their effects on longer dynamics.

The land managers of Eglin Air Force Base want to maintain or to increase the area of longleaf pine forest (Jackson Guard 1993). The alternative fire management strategies were evaluated based upon the change in area covered by either longleaf (good) or hardwood (bad) dominated sites. These changes were evaluated in terms of both the total area changed and the spatial pattern of vegetation on the landscape. By applying the same management strategies with different numbers of prescribed burns per year, the simulations were able to interpolate the minimum frequency of fire necessary for each strategy to maintain the total amount of longleaf pine forest. The changes in landscape pattern produced by alternative fire management strategies were compared by examining maps of the landscape.

Discussion of Burning Strategies

The simulation model revealed major differences among the three fire management strategies (i.e., wildfire, rotation, and responsive). In figure 9.4, the point at which the longleaf and hardwood/sand pine rates-of-change cross the x-axis indicates the fire frequency threshold at which the total area of good and poor vegetation is maintained. Wildfire is unable to maintain the current area of longleaf pine–dominated forest at either its current frequency or its historical frequency. The two prescribed-fire strategies perform better—both rotation and responsive strategies are able to maintain the area dominated by longleaf pine.

The number of prescribed fires set per year determines the transition rate between longleaf- and hardwood-dominated sites. With increased fire frequency comes increased conversion to longleaf pine, but the amount of fire necessary to maintain the existing area of longleaf pine differs substantially between these

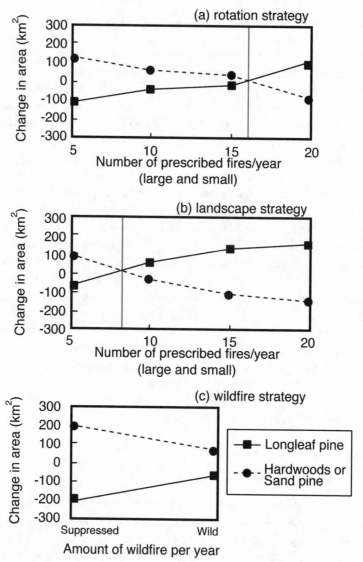

Figure 9.4. Changes in proportion of landscape covered by different vegetation classes in response to varying intensities of three management strategies: *(a)* a rotation management strategy, *(b)* a landscape or responsive management strategy, and *(c)* a wildfire management strategy. The independent axis represents prescribed burn intensity and ranges from five large and five small burns a year to twenty small and twenty large burns per year. The results shown are the average values of ten simulations. The standard errors are small, less than 0.5 percent of each mean value, and are not visible. The vertical lines in *(a)* and *(b)* indicate the minimum number of prescribed burns necessary to maintain the existing area of longleaf pine vegetation. Neither wildfire management strategy succeeded in maintaining the initial area of longleaf vegetation.

two prescribed-fire strategies. The responsive fire management strategy is able to maintain the existing area of longleaf pine by applying about eight small and large fires per year, which is equivalent to a fire frequency of 6.7 years per fire. The rotation strategy requires a fire frequency of about sixteen small and large fires per year, which is equivalent to a fire frequency of 3.5 years per fire.

Wildfire Management Strategy

Historically, wildfire was able to maintain most of the Eglin Air Force Base landscape in a longleaf pine–dominated state; however, it is unable to even maintain the current fragmented mosaic of longleaf pine let alone return it to its histori-

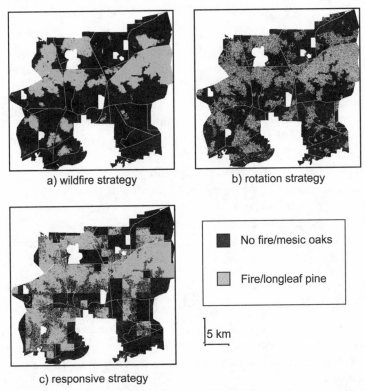

a) wildfire strategy

b) rotation strategy

c) responsive strategy

■ No fire/mesic oaks

▨ Fire/longleaf pine

5 km

Figure 9.5. Alternate fire-management strategies will change the spatial pattern of the landscape. Each map represents the landscape pattern at the end of a fifty-year simulation period. *(a)* The wildfire strategy would homogenize the landscape at small scales while dividing it into patches at large scales. *(b)* The rotation strategy would maintain the existing landscape, preserving heterogeneity across scales. *(c)* The responsive strategy would divide the landscape into patches at the landscape scale, but there is substantial diversity at local scales.

cal state. This change is due to the limited ability of wildfire to spread across areas of less-flammable vegetation. The fragmenting of the landscape into a set of isolated patches of longleaf pine limits the area that individual wildfires are able to burn, reducing the average size of forest fires and decreasing the frequency of fire experienced by sites within the fragmented landscape (figure 9.5a). Patches of longleaf pine must be larger than a specified minimum size to be ignited frequently enough to remain dominated by longleaf pine. The smaller the patch the higher the probability that it will not be ignited frequently enough to remain longleaf-pine dominated, and it will eventually be invaded by hardwoods and sand pine.

Rotation Fire Management Strategy

The rotation fire management strategy produces a landscape quite different from those produced by the other two strategies. Because it applies fire with equal frequency to all vegetation types across the landscape, this strategy tends to preserve existing landscape pattern. Under a wildfire strategy, the frequency with which a site is burned depends on both its vegetation and the vegetation that surrounds it. However, the rotation strategy burns sites independently of a site's vegetation and the vegetation that surrounds it. Consequently, the rotation strategy changes the spatial scales at which patterns are maintained within the landscape. Changes in fire frequency increase the dominance of hardwood or longleaf pine relatively evenly across the entire landscape. While wildfire homogenizes the landscape at the local level and divides the landscape into patches of longleaf pine–dominated and hardwood-dominated sites, the rotation strategy preserves the existing landscape pattern, decreasing heterogeneity at large spatial scales while preserving it at smaller scales (figure 9.5b).

Responsive Fire Management Strategy

The responsive fire management strategy produces a landscape with similarities to those produced by both the rotation and wildfire management strategies (figure 9.5c). The responsive strategy burns discrete blocks in the landscape, which causes it to produce small-scale heterogeneity similar to patterns produced by the rotation strategy. However, unlike the rotation strategy, the responsive strategy preferentially burns "good" areas, which contain high proportions of longleaf pine. This feedback between vegetation type and fire frequency produces a large-scale pattern of patches similar to that produced by wildfire. These patches of longleaf pine and hardwood forest even occur in locations similar to those produced by wildfire. This pattern arises because the responsive strategy (like the wildfire strategy) does not burn those areas with only small amounts of longleaf pine. Nonetheless, the responsive strategy with its rectangular prescribed burns imposes a geometric pattern of patches on the landscape that differs from the rounded patches produced by wildfire.

Spatial and Temporal Scales

The choice of different fire management strategies defines the spatial and temporal scales of fire. While vegetation succession is a local process that occurs over decades, prescribed and wildfires burn large areas and occur much more frequently. Succession shifts a forest site toward hardwood or sand pine dominance if fires occur less than every fourth year, while more-frequent fires move the forest toward a longleaf pine–dominated state. In a wildfire regime, the rate at which wildfires are initiated (number per area per year) defines the scale of the fire regime—it is the minimum area that will be ignited frequently enough to remain longleaf pine. Prescribed fires spread across a landscape in a controlled way, so the scales of fire regimes are defined by the size of burned areas and the area burned each year. The size of the prescribed fires determines the resolution scale at which prescribed fire alters the landscape. With the rotation strategy, the area of burn units and the total area burned per year define the frequency with which a site will be burned. If fire occurs less frequently than once every four years, then a site will move toward oak dominance; if fire occurs more frequently, then the site will move toward longleaf pine dominance. With the responsive strategy, however, some areas are burned more frequently than others. Thus, while the total area burned decreases, fire frequency in the burned areas increases—allowing sites to be more easily moved toward longleaf pine dominance.

The differences between a wild and a managed landscape can be shown by comparing the scales of structures and processes on a Stommel diagram. The critical fire frequency of four years per fire can be plotted as a straight horizontal line (figure 9.6). The rate of fire initiation (number per area per year) is plotted as a diagonal line (for any given initiation rate, fires are frequent at the large-scale but less frequent at smaller scales). For the wildfire strategy, the rate of fire initiation per area per year determines the minimum size of a patch of longleaf pine that can maintain itself as longleaf pine–dominated forest. This area can be approximated by the intersection of the fire frequency and the fire initiation rate on the Stommel diagram (figure 9.6a). The scale domain of a prescribed fire regime is determined by the spatial extent of burn area, the number of areas burned per year, and the total area being burned. This domain can be plotted on a Stommel diagram (figure 9.6b). The size of individual prescribed fires determines the resolution at which prescribed fire alters the landscape.

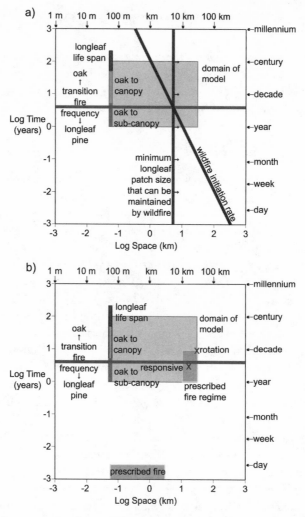

Figure 9.6. A comparison of the spatial and temporal scales of ecological and management processes that determine forest composition. Vegetation succession is a local process that occurs over long time periods and is shown in the upper left quadrant. Fires (both prescribed and wild) are relatively fast processes that occur over large areas and are shown in the lower left quadrant. (*a*) A fire frequency of about four years separate oak-dominated sites from longleaf-pine-dominated sites (horizontal line in diagram). The intersection of this frequency line with the rate of fire initiation per area per year line (diagonal line) identifies the minimum area that will be ignited frequently enough to remain longleaf pine. (*b*) When a landscape is burned by prescribed fire, the scale domain of fire is defined by the size of the prescribed fire units and the number of fires burned each year, which in turn determines the scale at which prescribed fire alters the landscape. The number of fires burned defines the total area burned each year area, while the total area being burned defines the fire frequency at which sites are burned. A rotation strategy may burn the entire landscape at a frequency that moves it toward oak dominance. However, if some sites are effectively burned more frequently than others, a portion of sites can be moved toward longleaf pine dominance.

In summary, fire management alters the relationship among ecological processes, eliminating the local connection (vegetation succession) among sites, and replacing it with connection at the scale of burn units. In the absence of prescribed fire, the interaction of wildfire and vegetation dynamics produces sharp edges between alternate stable states. Fire management policies can either amplify or attenuate these alternate states. The rotation strategy attenuates the forces producing these states: it decouples the burning of an area from its vegetation by attempting to burn all sites at the same frequency, regardless of their vegetative state. The result is a mixing of the vegetative states. The responsive strategy, like wildfire, connects fire and vegetation, amplifying the forces producing alternate stable states. However, it changes the scales at which succession and fire interact.

Managing Landscapes Containing Alternative Stable States

The model developed for Eglin Air Force Base suggests several principles that are generally applicable to the management of landscapes comprising a mosaic of alternative states. It suggests that the existence of alternative stable states can impede restoration and increases the importance of landscape pattern.

Limits to Restoration

The evaluation of wildfire restoration at Eglin Air Force Base demonstrates that the existence of alternative states presents barriers to restoration, especially when attempting to restore a landscape using a spatial process. Historically, wildfire has maintained relatively homogeneous longleaf pine savannas. However, restoring the historical fire regime will not restore the landscape to its original state, because of the reduced effect of fire on the fire-inhibiting alternative stable state hardwood-dominated forest and the reduced ability of fire to spread across the landscape. These changes in the forest reduce the average size of fires and decrease the ability of fire to burn in many areas. Consequently, restoring the historic fire regime exposes sites in the landscape to the same frequency of fire they experienced historically.

It is possible to make the generalization that as difficult as it is to restore an ecosystem after it reaches an undesirable state, this difficulty can be compounded when spatial processes maintain alternative states. If spatial connectedness amplifies the local forces producing the alternative states, then changes in the presence of one state in the landscape make it more difficult to return to previous patterns. The role of landscape pattern in changing ecological dynamics has been explored in many contexts in landscape ecology (Turner 1989; Perry

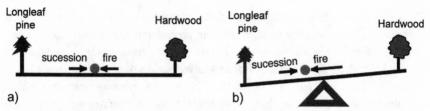

Figure 9.7. Alternate models of forest dynamics. *(a)* The *balance-of-forces model* of forest dynamics that underlies the rotation strategy. *(b)* The *teeter-totter model* of forest dynamics that underlies the responsive strategy.

1995). In this situation, the aggregation of longleaf pine and hardwood dominated sites into patches amplifies the forces driving particular sites toward either longleaf pine or hardwood dominance.

Recognizing Alternate Stable States

The presence of alternative states requires changes in management strategies, which often make implicit assumptions about the existence of alternative states that have altered their effectiveness. For example, at Eglin Air Force Base the rotation fire management strategy implicitly assumes that forest composition is determined by the relative balance between fire and succession (i.e., hardwood or sand pine invasion of longleaf pine). Consequently, this model suggests that by varying fire frequency one can obtain any desired forest composition (figure 9.7a). The responsive fire management strategy, on the other hand, presumes the existence of alternative stable states and assumes that fire changes hardwood sites less often than it changes longleaf pine sites. This model suggests that a given amount of fire will effect greater vegetation change if it is applied to longleaf pine rather than to hardwoods (figure 9.7b). Consequently, when the use of fire is limited or restricted, what is available can be used to greatest effect by applying it to areas with larger proportions of longleaf pine forest. One can conclude from this example that management strategies should allocate resources based upon their effectiveness in altering undesired states or maintaining desired states.

Importance of Burn Size

The costs of burning are largely independent of the size of the area burned, which suggests that the most cost-effective way to manage a landscape is to increase the average burn size whenever possible. Within a given year, for instance, the largest burn units should be burned first in order to maximize the total area burned each year. Over the long term, management practices that include not only prescribed fire but also other activities, such as tree planting, tree harvesting, and changes in land use practices, should be implemented so as to allow the largest possible areas to be burned at a given time.

Key Areas for Intervention

Landscapes composed of a mosaic of alternate stable states will frequently contain key areas that act to either fragment or connect areas in the same alternate state. Such areas are potentially key areas for management intervention. What defines these areas, however, will depend upon the ecosystem being analyzed; in general, though, their existence will make management of landscapes exhibiting alternate states more complex than that of simpler landscapes. Although this complexity is a barrier to some types of change, it will also offer opportunities for innovative management practices that use landscape pattern and the internal dynamics of alternative states to amplify management activities. The challenge for managing such landscapes is discovering and applying such policies.

Concluding Remarks

In northern Florida, forests can be dominated by either fire-encouraging longleaf pine or by fire-discouraging hardwoods or sand pine. Self-reinforcement of these forest types and their respective fire regimes leads to either fire-dominated (longleaf pine) or fire-excluded (hardwood and sand pine) states. Fire suppression and fragmentation have reduced areas of longleaf pine forest, and this loss has led land managers to attempt to restore previous areas of longleaf pine.

Fire management policies can either amplify or attenuate the strength of the forces driving the forest toward longleaf pine or hardwood dominance. A rotation strategy that burns each site on the landscape with a set periodicity attenuates these forces, stabilizing the relatively undesirable landscape. A responsive strategy, which selectively burns longleaf pine–dominated areas, amplifies the forces that create the alternate stable states. By amplifying the forests' natural dynamics, the responsive strategy can more effectively use fire to restore longleaf pine forests. However, this strategy, like most forms of management, changes the scales at which both ecological pattern and dynamics occur. The management of landscapes composed of a mosaic of alternate stable states presents challenges not found when managing simpler landscapes. But the case of Eglin Air Force Base suggests that crafting management practices based upon landscape dynamics can produce more effective management strategies.

Literature Cited

Abrahamson, W. G., and D. C. Hartnett. 1990. Pine flatwoods and dry prairies. Pp. 103–149 in *Ecosystems of Florida*, edited by R. L. Myers and J. J. Ewel. Gainsville: University Press of Florida.

Boyer, W. D., and D. W. Peterson. 1983. Longleaf pine. Pp. 153–156 in *Silvicultural systems for the major forest types of the United States*. U.S. Department of Agriculture, Washington, D.C.

Glitzenstein, J. S., W. J. Platt, and D. R. Streng. 1995. Effects of fire regime and habitat on tree dynamics in north Florida longleaf pine savannas. *Ecological Monographs* 65:441–476.

Hardesty, J., J. Adams, D. Gordon, and L. Provencher. 2000. Simulating management with models. *Conservation Biology in Practice* 1:26–31.

Heyward, F. 1939. The relation of fire to stand composition of longleaf pine forests. *Ecology* 20:287–304.

Holling, C. S. 1996. Engineering resilience versus ecological resilience. Pp. 31–43 in *Engineering within ecological constraints*, edited by P. C. Schulze. Washington, D.C.: National Academy Press.

Jackson Guard (Eglin Air Force Base). 1993. *Natural resources management plan*. Eglin Air Force Base, Fla.

Parker, K. C., and J. L. Hamrick. 1996. Genetic variation in sand pine (*Pinus clausa*). *Canadian Journal of Forest Resources* 26:244–254.

Perry, D. A. 1995. Self-organizing systems across scales. *Trends in Ecology and Evolution* 10:241–244.

Peterson, G. D. 1999. Contagious disturbance and ecological resilience. Ph.D. diss. Gainesville: University of Florida.

Platt, W. J., G. W. Evans, and S. L. Rathburn. 1988. The population dynamics of a long-lived conifer (*Pinus palustris*). *American Naturalist* 131:491–525.

Platt, W. J., and S. L. Rathburn. 1993. Dynamics of an old-growth longleaf pine population. Pp. 275–297 in *Tall timbers fire ecology*, edited by S. M. Hermann. Tallahassee, Fla.: Tall Timbers Research Station.

Provencher, L., K. E. M. Galley, B. J. Herring, J. Sheehan, N. M. Gobris, D. R. Gordon, G. W. Tanner, J. L. Hardesty, H. L. Rodgers, J. P. McAdoo, M. N. Northrup, S. J. McAdoo, and L. A. Brennan. 1998. *Post-treatment analysis of restoration effects on soils, plants, arthropods, and birds in sandhill systems at Eglin Air Force Base, Florida*. Annual report to Natural Resource Division, Eglin Air Force Base, Niceville, Fla; Public Lands Program, The Nature Conservancy, Gainesville, Fla.

Rebertus, A. J., G. B. Williamson, and E. B. Moser. 1989. Longleaf pine pyrogenicity and turkey oak mortality in Florida xeric sandhills. *Ecology* 70:60–70.

Robbins, L. E., and R. L. Myers. 1992. *Seasonal effects of prescribed burning in Florida: A review*. Miscellaneous Publication 8, Tallahassee, Fla.: Tall Timbers Research Station.

Turner, M. 1989. Landscape Ecology: The effect of pattern on process. *Annual Review of Ecology and Systematics* 20:171–197

PART III
Summary

10

A Summary and Synthesis of Resilience in Large-Scale Systems

Lance H. Gunderson, Lowell Pritchard Jr.,
C. S. Holling, Carl Folke, and Garry D. Peterson

We attempt four tasks in this the final chapter. We begin in this section with a set of paradoxes that have emerged from the case studies discussed in this volume (and elsewhere) to help highlight tensions and clarify arguments. The second section summarizes the information presented in the case studies by evaluating a set of propositions relating to resilience. We use the propositions presented in the first chapter and the theoretical refinements of the second chapter as a framework for that summary. The third section builds a broader understanding of resilience by contrasting the multiple meanings of the concept and identifying processes that change the resilience of a system. As a final task, we attempt in the fourth section to summarize the implications for managing these large, complex systems and conjecture whether one can indeed manage for resilience.

We present in this section a set of paradoxes that arise from a review of patterns chronicled in this volume as well as from other experiences and analyses of large, regional-scale ecosystems. In each of these systems, humans interact with the ecologic components in a variety of ways, for instance, by directly removing resources, controlling key processes, and depositing materials from other systems, as well as many other interventions. We will use the following paradoxes as an entrée into a deeper discussion of the behavior of these coupled systems of people and nature. We propose resolutions to these paradoxes following a brief discussion.

The Holling Frustration: The Pathology of Constancy Versus the Viability of Variability

For at least two decades, one of us (Holling 1978) has described a recurring pattern of pathology in the management of large-scale systems. Managers of natural resources systems are often successful at rapidly achieving a set of narrowly defined goals. Those goals are intended for the most part to control unwanted variation in the resource system. Examples include spraying to control spruce budworm in boreal forests (Baskerville 1995) and designing water control systems to constrict variability of water levels in the Everglades (Light et al. 1995) and variability of salinity in Florida Bay (chapter 6). The success in controlling one process encourages people to build up a dependence on continued control while simultaneously eroding ecological support. This leads to a state in which ecological change is increasingly undesirable to the people dependent upon the natural resource, and more difficult to avoid—such as those in the semi-arid rangeland examples of Walker (chapter 7). In these cases, the loss of resilience was due to constraint of the variability (in time and space) of key ecosystem processes. The erosion of resilience was manifest by the sudden switch in ecosystem structure, triggered by a disturbance that the system would have previously absorbed.

The Bite-Back Paradox: Well-Behaved Functions Versus Shifting Controls

This paradox is named for a colloquial expression that describes a situation where an intended design or solution produces an unintended consequence (Tenner 1996). This paradox highlights the tension between predictability and unpredictability, and it provides an analytic framework for understanding complex dynamics. Ludwig et al. (chapter 2) show the unpredictability of system dynamics using just three nonlinear coupled equations. Many of the case studies (Carpenter and Cottingham chapter 3; Jansson and Jansson chapter 4; Walker chapter 7) indicate the results of shifting controls and corresponding surprising ecosystem responses. The short answer to this paradox is that well-behaved functions are not so well behaved over time.

The MacArthur Paradox: Diversity Increases Stability Versus Diversity Increases Resilience

Over the past three decades, ecological work has focused on the relationship between biodiversity and ecosystem function. One group has demonstrated the relationship between diversity and stability (e.g., recovery) of ecosystem function (Tilman and Downing 1994; Tilman et al. 1996). Another group's work demonstrates the relationship between diversity and resilience (Walker 1992, 1995; Walker et al. 1997, 1999; Peterson et al. 1998). The resolution to the paradox is that it is not an either/or proposition; indeed, stabilizing properties coexist with resilient properties. We attempt some clarification of this duality in following sections.

The Sustainability Paradox: Short-Term Efficiency Versus Long-Term Sustainability

Many of the arguments regarding sustainability focus on future options. Those options can be specific obligations by one generation to subsequent ones, such as endangered species preservation or a more generic responsibility of maintaining a diversity of options such as those suggested by biodiversity arguments (e.g., biological diversity should be preserved for undiscovered medicinal uses, for ecological resilience or for biophilia-type arguments). Yet, large portions of human activities are designed for efficient use of resources. The concepts of ecological resilience and institutional flexibility can help reconcile the paradox between short-term efficient use of resources and long-term sustainability. We begin that discussion of resilience by reviewing and evaluating propositions posed earlier in this volume.

Evaluation of Propositions

In chapter 1, a set of propositions on resilience in large-scale systems was presented. In this section, we will evaluate those propositions. Evaluation will be made through inference rather than through a formal process of null rejection. The case studies provide information that can be used to suggest refinements, clarification, and revisions of the propositions. That is, the propositions are useful heuristics around which one can organize thoughts and ideas from observations and accumulated knowledge of the dynamics and transitions of complex ecosystems. We begin with a statement of each proposition, followed by a discussion that assesses each in terms of the information from the case studies.

The Organization of Regional Resource Systems Emerges from the Interaction of a Few Variables

The essential structure and dynamics of complex systems are produced by the interaction of at least three and no more than six variables. These variables operate at scales that differ, both spatially and temporally, by approximately an order of magnitude. In constructing models, this has been described as the "Rule of Hand" because the number of variables should be less than the number of fingers on one hand (Holling et al. 2002). It is not a mathematical theorem; rather, it is a loose guideline for developing ecological models that are complex enough to generate multiple-state behavior (chapter 2; chapter 3), but simple enough to understand. The components of the model exhibit at least three quite different (by around ten-fold) turnover times. Some of the linkages among components are nonlinear in the state variables (but not necessarily in the parameters). The slow variables in these models create a stability landscape with multiple attractors. Because the slow variables are dynamic, and have feedbacks with the fast variables, the stability landscape is itself dynamic, so transitions among attractors are possible.

Complex Systems Have Multiple Stable States

Complex systems can exhibit alternative stable organizations. Transitions between different organizations are due to changes in the interaction of structuring variables. Change often occurs when gradual change in a slow variable alters the interactions and relationships among faster variables.

Since the introduction of the concept of multiple stable states into the ecological literature (Holling 1973), only a handful of references have addressed the key point of whether these states exist. Clearly, the states are demonstrable from modeled ecosystems (Ludwig et al. 1978; Carpenter et al. 1999; chapters 2 and 3). The aim of this volume was to demonstrate the presence of multiple stable states in a wide variety of systems and describe those states in ecological terms.

This volume provides myriad examples of the presence of alternative stable states in ecosystems. Those examples cross biomes and different structural complexities. Multiple states are demonstrated for aquatic systems such as inland freshwater lakes, inland seas, and coral reefs, wetland ecosystems, and terrestrial systems of rangelands, and tropical and temperate forests.

Alternative stable states in the systems dominated by water (aquatic and wetland systems) are characterized by shifts in pathways of energy flow and trophic structure. The sources of primary productivity in freshwater lakes, inland seas, and shallow bays changed from benthic vegetation to phytoplankton in the water column (chapters 3 and 6). In the coral reef systems and lake systems, these shifts in primary production were followed by shifts in trophic structure, as populations of keystone species were either enhanced or decreased.

The alternative states in terrestrial systems were manifest as changes in species composition and species dominance. The rangeland systems undergo changes from grass dominance to woody shrub dominance. In the forest systems, dominance of both overstory and understory species shift with the change in stable state.

Resilience Derives from Functional Reinforcement across Scales and from Functional Overlap within Scales

Resilience derives from the combination of a replication of function across a range of spatial and temporal scales and from a diversity of similar, but different, functions operating within a scale. These functions arise from the interaction of a diverse set of biotic and physical processes that control net carbon assimilation and transpiration, water extraction from various soil layers, nutrient cycling and retention, herbivory, and predation. These processes involve chemical and physical ones interacting with processes mediated by critical species in the biota.

These species can be divided into functional groups based upon differences in their ecological functions. For plants, these different functions are represented by attributes such as nitrogen fixing capacity, rooting depth, water-use efficiency, and litter decomposition rate. For animals, they are represented by trophic status, body mass, and foraging class.

Adequate performance of ecosystem function depends on having all the necessary functional groups present (the full array, or diversity, of functional groups). The persistence of ecosystem function over time (i.e., the resilience of ecosystem function) depends upon the diversity of species within functional groups. There are two important forms of diversity within functional groups: one provides functional compensation within a narrow range of scales and one provides functional reinforcement across a wide range of scales (Peterson et al. 1998; Holling et al. 2002).

Functional compensation within a narrow range of scales occurs when species perform a similar function but have different environmental sensitivities. A study of functional attribute diversity of an Australian rangeland (Walker et al. 1999) revealed that the most abundant plant species were far apart from each other in plant attribute space (i.e., they perform different functions). However, among the less-abundant species, there was at least one species functionally very similar to each abundant species. Furthermore, on a site that had been heavily grazed, dominant species that had been eliminated were replaced by one of their functionally similar species that were among the less-abundant species on the lightly grazed site. It is an example of resilience achieved from functional compensation within a scale range.

Vulnerability Increases As Sources of Novelty Are Eliminated and Functional Diversity and Cross-Scale Functional Replication Are Reduced

Diminished sources of novelty reduce the ability of a system to recover from disturbances. The elimination of structuring species or processes can cause an ecosystem to reorganize, while a reduction in functional diversity and replication reduces the ability of functions, and therefore of a system, to persist.

The propositions evaluated in this volume are a beginning toward understanding how ecosystems change over time and space with and within human intervention. All are based on observations and precepts that suggest nonlinear dynamics, indicative of shifting ecosystem controls. We explore those dramatic shifts in controls and states in the next section on ecological resilience.

Understanding Ecological Resilience

Resilience is about characterizing and understanding change in complex ecosystems. Most of the cases in this volume focus on how human intervention results in ecosystem state change. A change in ecosystem state is indicative of a switch in the small set of controlling processes in the ecosystem. To help understand this dynamic, the remainder of this section is structured around definitional problems of resilience, mechanisms that generate state change, and processes that either add to or detract from ecosystem resilience.

Multiple Meanings

The first two chapters presented both qualitative and quantitative frameworks that can be used to help understand multiple meanings associated with resilience. The subsequent chapters use a case-study format within those frameworks to demonstrate and in some cases enrich the variation of meanings around resilience—including return time, state shifts, hysteresis, reversibility, adaptive capacity, and scales.

Engineering Resilience Versus Ecological Resilience

Two contrasting conceptions of resilience were introduced in chapter 1. Ludwig, Walker, and Holling (chapter 2) demonstrated the mathematics behind those who use resilience to mean the time required for a system to return to an equilibrium or steady state following a perturbation (Appendix, chapter 2). Implicit in this definition is the assumption that the system exists near a single or global equilibrium condition. Holling (1996) describes this form as "engineering resilience," while others consider it a property of "stability" (for example Ludwig et al. in chapter 2 treat how far the system has moved from an equilibrium in time and how quickly it returns as a measure of stability). Other fields that use the term resilience, such as physics, control system design, and material engineering, all use this definition of resilience.

The second type of resilience (called "ecological resilience" in chapter 1 and in Holling 1996) is measured by the magnitude of disturbance that can be absorbed before the system redefines its structure by changing the variables and processes that control behavior (Holling 1973; Ludwig et al. chapter 2). As noted in chapter 1, the concept of ecological resilience presumes the existence of multiple stability domains and the tolerance of the system to perturbations that facilitate transitions among stable states. Hence, ecological resilience refers to the width or limit of a stability domain and is defined by the magnitude of disturbance that a system can absorb before it changes stable states.

Carpenter and Cottingham (chapter 3), Carpenter et al. (1999), and Scheffer (1998) have used the heuristic of a ball and a cup to highlight differences between these types of resilience. The ball represents the system state and the cup represents the stability domain (figure 10.1). The ball sitting at the bottom of the cup depicts equilibrium. Disturbances (depicted by an arrow) move the

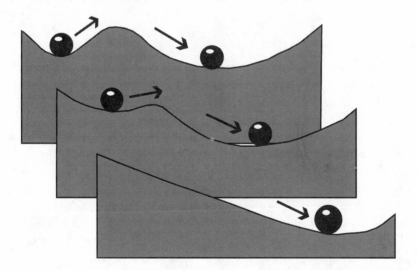

Figure 10.1. Three-dimensional heuristic indicating stability, resilience, and adaptive capacity. Ecological resilience is exceeded when a disturbance causes a shift in stability domains (moves over a hump in the landscape), as shown in the two back slices. However, the stability landscape changes (succession in the fire examples, or nutrients in the lake and wetland examples) as shown by the three slices. The ability of the system to maintain structure and function with a changing landscape is called *adaptive capacity*.

marble to a transient position within the cup. Engineering resilience refers to characteristics of the shape of the cup—the slope of the sides dictate the return time of the ball to the bottom. Ecological resilience suggests that more than one cup exists, and resilience is defined as the width at the top of the cup. Implicit in both of these definitions is the assumption that resilience is a static property of systems. That is, once defined, the shape of the cup remains fixed over time. However, the cases in this volume indicate that stability domains are dynamic and variable.

Adaptive Capacity

Many of the manifestations of human-induced state changes in ecosystems result from alteration of the key variables that influence the underlying stability domains. The key variables that configure these stability domains change at relatively slow rates without human intervention and change at more rapid rates with human intervention. The mechanisms for these shifts are elaborated in the next section. Using the ball-in-cup heuristic, the shape of the cup is subject to change, altering both stability (return time) and resilience (width of stability domain). Scheffer et al. (1993) depict this as multiple stability landscapes (three slices in figure 10.1). The property of an ecosystem that describes this change in stability landscapes and resilience is referred to as "adaptive capacity" (Peterson et al. 1998; Gunderson 2000).

Hysteresis and Reversibility

Ludwig and colleagues (chapter 2) make a distinction between hard and soft losses of resilience, which suggests that some state changes are reversible while others are not. Although reversibility of state changes implies that once a system changes state, it remains in that state indefinitely, some of the case studies such as the freshwater lakes, Florida Bay, temperate pine forests, and rangelands, appear reversible (or at least reversible within human time frames). The likelihood of reversible state change is less obvious in other cases (nutrient enrichment in wetlands and seas, coral reef degradation). The case of rangeland resilience has perhaps one of the longest terms of observation; indeed Walker published on state changes decades ago (Walker et al. 1981). The rangelands exhibit hysteresis; that is, the system returns to a prior state, but by means of a different path of recovery (Walker chapter 7; Ludwig et al. chapter 2).

Scales and Dimensions of Resilience

The multiple meanings of resilience all relate to or are dependent upon a time domain. The first definition (Holling 1973) was based on system dynamics over time, as are the more recent treatments (Ludwig et al. chapter 2; Carpenter et al. 1999; Gunderson 2000). The chapters by McClanahan, Polunin, and Done, (chapter 5) and Peterson (chapter 9) introduce the spatial dimension of resilience. This is manifest as a transition from fixing the spatial bounds of a system for analysis to one in which the spatial dimensions are integrated with the temporal ones to yield even more dynamic systems, as Peterson demonstrates (chapter 9).

Ecological resilience and adaptive capacity both suggest interactions across scales. One conclusion from these treatments is that resilience can only be discussed, analyzed, and measured across scale ranges (either in space or time or both). These themes are developed in the next section, which describes how human interventions result in state changes through the reduction of ecosystem resilience.

Loss of Ecological Resilience

The cases in this volume and a growing body of literature document transitions among stability domains in a variety of ecosystems (Carpenter and Kitchell 1993; Hanski 1995; Ludwig et al. 1997; Weaver et al. 1996). Many of those systems are influenced by human activities, which has led to a confounding problem around ecological resilience. Sousa and Connell (1983) suggest that alternative stable states do not exist in systems untouched by humans, while others (Dublin et al. 1990; Sinclair et al. 1990) indicate that these types of dynamics are and have been part of systems both with and without humans. Without touching on the question of whether people are or are not natural parts of ecosystems, we see that people do change the resilience of systems. Human interventions and practices decrease ecological resilience through at least four processes: mining, eutrophication, modifying key ecosystem relationships, and homogenizing temporal and spatial variability. Each is discussed in turn.

In many of the systems, humans remove or extract key elements of the system. When these extraction rates exceed inputs, the net result is mining of ecosystem capital. In this sense, ecosystem capital is used as a general term to describe attributes such as species richness or biodiversity, stocks of nutrients, biomass, or organic matter. The loss of key species in three of the ecosystems resulted in a change of system state following disturbances such as droughts, fires, or hurricanes that were previously absorbed. In the rangelands, grazing removed drought-tolerant plants in favor of less-resistant species. When droughts occurred, the intolerant plants died, opening up space for shrubs to become established (Walker chapter 7). In the tropical forests of Puerto Rico, the key capital eroded by human activity was the organic soil and the multiple sources of regeneration in that soil (Lugo et al. chapter 8).

In other systems, primarily aquatic or wetland, run-off from human activities resulted in eutrophication (chapters 3, 4, 5, and 6). Increased run-off of nitrogen and phosphorus from agricultural activities accelerated the process of eutrophication in the freshwater lakes (Carpenter and Cottingham chapter 3) and oligotrophic wetlands (Gunderson and Walters chapter 6). At larger scales, nutrient run-off into the Baltic Sea increased areas of anoxia and caused dramatic shifts in flora and fauna (Jansson and Jansson chapter 4). Similar situations are discussed for the shift in coral reef structure, although eutrophication effects are confounded with other factors (McClanahan et al. chapter 5).

The third category of human intervention that appears to decrease ecosystem resilience is the modification of key ecosystem linkages or relationships. These can be trophic relationships, such as those demonstrated for freshwater lakes and coral reefs, where humans reduce populations of keystone fish species. Those changes in populations alter predation rates on other populations, leading to dramatic increases, such as the trophic cascade described by Carpenter and Kitchell (1993). Outbreaks of urchins on coral reefs that degrade reef structure can be linked to overfishing of key predators that control urchin populations (chapter 5). Functional relationships can also be modified by human interventions, as suggested by the coral reef chapter and the chapter on southern forests.

One of the more interesting but perhaps least explored changes has to do with homogeneous spatial patterns. The loss of resilience in Florida Bay (chapter 6) was attributed to a spatially homogeneous seagrass biomass. When oxygen in the water column decreased, respiratory demands were too great for photosynthesis, and the individual plants began to die. The die-off spread to larger scales, as dead plant material removed more oxygen from the water and shaded other plants. The removal of small-scale disturbance agents such as turtles and manatees is thought to have played a role in the resulting even biomass distribution. Similar arguments are put forth in the chapter on southern forests, where the spatial homogeneity of stands has changed the size and severity of fires in the landscape (Peterson chapter 9).

In ecological systems, resilience lies in the requisite variety of functional groups and the accumulated capital that provides sources for recovery. Resilience within a system is generated by destroying and renewing systems at smaller, faster scales. Ecological resilience is reestablished by the processes that contribute to system "memory"—those involved in regeneration and renewal that connect that system's present to its past and to its neighbors. Management regimes or human activities that remove variability in either space or time will decrease system resilience. Addition of nutrients directly or indirectly leads to changes in abiotic substrates and subsequent irreversibility in the biology.

Management and Resilience

When shifts occur between alternative states or conditions, it is usually signaled as a resource crisis. That is, a crisis occurs when an ecosystem behaves in a surprising manner or when observations of a system are qualitatively different from peoples' expectations of that system. Such surprises occur when variation in broad-scale processes (such as a hurricane or extreme drought), intersects with internal changes in an ecosystem due to human alteration. Examples from this volume include the emergence of shrubs in semi-arid rangelands (Walker chapter 7), algae blooms in freshwater lakes (Carpenter and Cottingham chapter 3), and shifts in vegetation due to nutrient enrichment in the freshwater Everglades (Gunderson and Walters chapter 6). With each of these shifts in stability domains chronicled as crisis, understanding how and why people chose to react is key to managing for resilience.

When faced with shifting stability domains and corresponding crises, management options fall into one of three general classes of response. The first is to do nothing and wait to see if the system will return to some acceptable state while sacrificing lost benefits of the undesirable state. The second option is to actively manage the system and try to return the system to a desirable stability domain. The third option is to admit that the system is irreversibly changed, and hence the only strategy is to adapt to the new, altered system. The resilience of the system provides the ability to cope or adapt in a world characterized by crises and shifting stability domains. We build upon this theme in the following two sections, one on how building understanding provides resilience and the second on how to maintain or restore resilience in managed systems.

Uncertainty, Understanding, and Resilience

In this section, we develop a set of arguments that are based on the proposition that confronting uncertainty through learning and understanding contributes to resilience. How people choose to deal with uncertainty is key to either increasing or losing resilience. Effective responses are those that assess types and sources of uncertainties, but also identify sources of flexibility, as well as develop actions that are structured for learning and allow for the generation of novelty.

During most of this century, the goal of technologically based resource management has been to control the external sources of variability in order to seek a singular goal, such as maximization of yield (trees, fish) or controlling levels of pollution. This approach, also called "command and control," focuses on controlling a target variable, which is successful at first, but then slowly changes other parts of the system. That is, by isolating and controlling the variables of interest (i.e., assuming that the uncertainty of nature can be replaced with the certainty of control) we have as a result eroded ecosystem resilience. The manifestation of that erosion is the pattern of policy crisis and reformation as mentioned above and elsewhere (Gunderson et al. 1995). Much of subsidized agriculture, where incentives are set up to deal with changes in markets and costs, as well as surprises from nature, falls into this category.

People involved in the practice of resource management are all linked by the need for understanding. The experience and practice during this century has been to turn to scientists, who are the heart and soul of technology, and technologic solutions as the fountains of understanding. But there has been a growing sense that traditional scientific approaches are not working and indeed are making the problem worse (Ludwig et al. 1993). Two reasons why rigid scientific and technological approaches fail is because they tend to focus on the wrong types of uncertainty and on narrow types of scientific practice. Many formal techniques of assessment and policy analysis presume a system near equilibrium, with a constancy of relationships, and that uncertainties arise not from errors in tools or models, but from lack of appropriate information that go into the models. Often, computer models are used to integrate understanding, which when properly transformed provide resilience in the form of options that acknowledge the uncertainties of the issue.

A conflict also arises between two views of science that contributes to the loss of resilience. One mode of science focuses on parts of the system and deals with experiments that narrow uncertainty to the point of acceptance by peers; it is conservative and unambiguous by being incomplete and fragmentary. The other view is integrative and holistic, searching for simple structures and relationships that explain much of nature's complexity. This view provides the underpinnings for an approach to dealing with resource issues called *adaptive management*, which assumes surprises are inevitable, that knowledge will always be incomplete, and that human interaction with ecosystems will always be evolving.

Adaptive management is an integrated, multidisciplinary method for natural resources management (Holling 1978; Walters 1986). It is adaptive because it acknowledges that the natural resources being managed will always change and therefore humans must respond by adjusting and conforming as situations change. There is and always will be uncertainty and unpredictability in managed ecosystems, both as humans experience new situations and as these systems change as a result of management. Surprises are inevitable. Active learning is the way in which this uncertainty is winnowed. Adaptive management acknowledges that policies must satisfy social objectives but must also be continually modified and flexible for adaptation to these surprises. Adaptive management therefore views policy as hypotheses—that is, most policies are really questions masquerading as answers (Light pers. comm.). Since policies are questions, then management actions become treatments in the experimental sense. The process of adaptive management includes highlighting uncertainties, developing and evaluating hypotheses around a set of desired system outcomes, and structuring actions to evaluate or "test" these ideas. Although learning occurs regardless of the management approach, adaptive management is structured to make that learning more efficient. Trial and error is a default model for learning while managing; people are going to learn and adapt by the simple process of experience. Just as the scientific method promotes efficient learning through articulating hypotheses and putting those hypotheses at risk through a test, adaptive management proposes a similar structure.

A unique property of human systems in response to uncertainty is the generation of novelty. Novelty is key to dealing with surprises or crises. Humans are unique in that they create novelty that transforms the future over multiple decades to centuries. Natural evolutionary processes cause the same magnitude of transformation over time spans of millennia. Examples are the creation of new types and arrangements of management institutions after resource crises in the Everglades (Light et al. 1995), Columbia River basin (Lee 1993) or the Baltic Sea (Jansson and Velner 1995). In technologies, invention and adaptations transform the future (Arthur et al. 1987; Kauffman 1995).

Restoration and Maintenance of Resilience

In this final section, we seek to develop concepts regarding the maintenance or restoration of resilience in resource systems. At least two aspects can be identified: strategies that people employ in order to manage for resilience in a resource system and properties that contribute to resilience in human organizations. In order to add resilience to managed systems, a number of strategies are employed: increase the buffering capacity of the system, manage for processes at multiple scales, and nurture sources of renewal. We begin with the role of buffering to manage for resilience in various systems.

Most activities for buffering tend to address the engineering type of resilience, that is, mitigating the effects of unwanted variation in the system in order to facilitate a return time to a desired equilibrium. In many agricultural systems, resistance to change is dealt with by a combination of barriers to outside forces (tariffs, fences, etc.) and internal adjustments such as water cost control mechanisms (Conway 1993). Water resource systems can be designed for resilience by increasing the buffering capacity or robustness through redundancy of structures (and flexibility of operations) rather than by employing fewer, larger structures and rigid operational schemes (Fiering 1982). Berkes and Folke (1998) suggest that traditional approaches (which they define as traditional ecological knowledge) buffer managed systems by not allowing unpredictable or large perturbations to threaten ecosystem structure and function by allowing smaller-scale perturbations to enter the system. One such example is that of the Cree fishermen in northern Canada, who use a mixed-size mesh net to harvest multiple age classes, thereby preserving an age class structure that mimics a natural population (Berkes 1998). This stable age structure helps buffer widely varying reproductive success. Closely related to buffering is the notion of resilience in economic systems.

In economic systems, multiple technologies add resilience in the face of shifts in demand and factor prices and availability, but maintaining multiple technologies is costly. With stable (or artificially stabilized) demand, a firm or an industry, lacking the need for flexibility, could focus its attention on the incremental gains in efficiency available from moving to and refining the "best technology." This is equivalent to a dynamic shift to the low point on the long-run average cost curve and the synchronization of technology across firms. When conditions ultimately change, the resilience of the system depends inversely on the degree of specialization in capital and skills that were developed. Depending on the forward and backward linkages, the rest of the economy may also be adversely affected by the conditions in a keystone industry.

Resource systems that have been sustained over long time periods increase resilience by managing processes at multiple scales. Returning to the example of the Cree, Berkes (1995) argues that multiple spatial domains are part of their fishing practices and multiple temporal domains in their hunting. While fishing, the Cree monitor catch per unit effort. When they notice the rate dropping, they immediately move to alternative fishing sites within a geographically smaller area; over longer time frames, they rotate fishing effort to more remote sites (Berkes 1995). Similarly, they retain information through belief systems, for example, that caribou will return for hunting at annual and decadal cycles or periods.

Another example is in the Everglades water management system, where in the mid-1970s water deliveries to Everglades Park were based upon a seasonally variable but annually constant volume of water. This system was changed in the mid-1980s to a statistical formulation that incorporated interannual variation into the volumetric calculation (Light et al. 1995). Berkes and Folke (1998) argue that local communities and institutions coevolve by trial and error at time scales in tune to the key sets of processes that structure ecosystems within which the groups are embedded. Many of the crises chronicled in Gunderson et al. (1995) were created by an inherent focus on one scale for management and reformations of learning recognized the multiple scales by which the ecosystem was functioning.

Another way in which people manage for resilience in resource systems is by concentrating on sources of renewal. Many forms of catastrophic insurance provide this function by creating a fiscal reservoir that can be tapped should structures need to be replaced. Another mechanism that explicitly plans for renewal in resource systems is the scheme of market-based property rights systems developed for Australia (Young 1992). Young and McCay (1995) argue that adding flexibility and renewable structure to property rights regimes will increase resilience. They indicate that market-based property right schemes (licenses, leases, quotas, or permits) should have a built-in sunset (termination) to the scheme, with stable arrangements (entitlements, obligations) in the interim years. These principles complement Ostrom's (1990, 1995) findings that successful institutions allow stakeholders to participate in changing rules that affect them.

Such mechanisms highlight the profound political nature of managing for resilience. Economies are nested in political structures such as those institutions that secure property rights and add the force of law to contracts. Perhaps in the field of political economy it is the economy (the market—that is, the fast process) that manages for stability within a given set of entitlements and endowments (defined by political institutions—the slow variable), which are in turn justified by shared culture. It is in the political realm that choices between alternative social structures, divisions of income, and levels of governance are made, and it is there that resilience comes into question. Once a set of endowments is determined, the market can find an equilibrium that is stable (in the sense of Pareto-efficiency, which is the First Theorem of Welfare Economics). Whether that equilibrium is politically stable is another matter, and that depends on the resilience of the underlying sociopolitical structure.

Institutions (defined broadly as the set of rules, or structures, that allow people to organize for collective action) can add resilience to a system. A few key ingredients appear necessary to facilitate the movement of systems out of crisis through a reformation. In the review of management histories in western systems (Gunderson et al. 1995), these included functions of learning, and engagement and the ability to tap into deeper understanding and trust. Kai Lee calls this "social learning," by combining adaptive management frameworks within a framework of collective choice (1993). Other authors (Berkes and Folke 1998) describe this as social capital composed of the institutions, traditional knowledge, and common property systems that are the mechanisms by which people link to their environment. It is such linkages and connectivity across time and among people that helps navigate transitions through periods of uncertainty to restore resilience.

Concluding Remarks

Resilience in engineering systems is defined as a return time to a single, global equilibrium. Resilience in ecological systems is the amount of disturbance that a system can absorb without changing stability domains. Adaptive capacity refers to the ability of a system to persist with shifting stability domains. In ecological systems, resilience lies in the requisite variety of functional groups and the accumulated capital that provides sources for recovery. Resilience within a system is generated by destroying and renewing systems at smaller, faster scales. Ecological resilience is reestablished by the processes that contribute to system "memory," those involved in regeneration and renewal that connect that system's present to its past and it to its neighbors.

A resource crisis sometimes signals a shift in stability domains. When a system has shifted into an undesirable stability domain, the management alternatives are to (1) restore the system to a desirable domain, (2) allow the system to return to a desirable domain by itself, or (3) adapt to the changed system because changes are irreversible.

Resilience is maintained by focusing on (1) keystone structuring processes that cross scales, (2) sources of renewal and reformation, and (3) multiple sources of capital and skills. No single mechanism can guarantee maintenance of resilience. Strategies that address a requisite variety of purposes and concentrate on renewal contribute to resilience. Institutions should focus on learning and on the understanding of key cross-scale interactions. Learning, trust, and engagement are key components of social resilience. Social learning is facilitated by recognition of uncertainties, monitoring, and evaluation by stakeholders. The most difficult issues to deal with are those whose consequences will be realized ten to fifty years in the future over broad scales.

Literature Cited

Arthur, W. B., Y. M. Ermoliev, and Y. M. Kaniovski. 1987. Path-dependent processes and the emergence of macro-structure. *European Journal of Operations Research* 30:294–303.

Baskerville, G. L. 1995. The forestry problem: Adaptive lurches of renewal. Pp. 37–102 in *Barriers and bridges to the renewal of ecosystems and institutions*, edited by L. H. Gunderson, C. S. Holling, and S. S. Light. New York: Columbia University Press.

Berkes, F. 1998. Indigenous knowledge and resource management systems in the Canadian subarctic. Pp. 98–128 in *Linking social and ecological systems: Management practices and social mechanisms for building resilience*, edited by F. Berkes and C. Folke. Cambridge: Cambridge University Press.

Berkes, F., and C. Folke. 1998. Linking social and ecological systems for resilience and sustainability. Pp. 1–25 in *Linking social and ecological systems: Management practices and social mechanisms for building resilience*, edited by F. Berkes and C. Folke. Cambridge: Cambridge University Press.

Carpenter, S. R., and J. F. Kitchell, eds. 1993. *The trophic cascade in lakes*. Cambridge: Cambridge University Press.

Carpenter, S. R., D. Ludwig, and W. A. Brock. 1999. Management of eutrophication for lakes subject to potentially irreversible change. *Ecological Applications* 9:751–771.

Conway, G. 1993. Sustainable agriculture: The trade-offs with productivity, stability and equitability. Pp. 46–65 in *Economics and ecology: New frontiers and sustainable development*, edited by E. B. Barbier. London: Chapman and Hall.

Dublin, H. T., A. R. E. Sinclair, and J. McGlade. 1990. Elephants and fire as causes of multiple stable states in the Serengeti-Mara woodlands. *Journal of Animal Ecology* 59:1147–1164.

Fiering, M. B. 1982. Alternative indices of resilience. *Water Resources Research* 18:33–39.

Gunderson, L. H. 2000. Ecological resilience—in theory and application. *Annual Review of Ecology and Systematics* 31:425–439.

Gunderson L. H, C. S. Holling, and S. S. Light, eds. 1995. *Barriers and bridges to the renewal of ecosystems and institutions*. New York: Columbia University Press.

Hanski, I. 1995. Multiple equilibria in metapopulation dynamics. *Nature* 377:618–621.

Holling, C. S. 1973. Resilience and stability of ecological systems. *Annual Review of Ecology and Systematics* 4:1–23.

———. 1978. *Adaptive environmental assessment and management*. London: John Wiley and Sons.

———. 1996. Engineering resilience versus ecological resilience. Pp. 31–43 in *Engineering within ecological constraints*, edited by P. C. Schulze. Washington, D.C.: National Academy Press.

Holling, C. S., L. H. Gunderson, and G. D. Peterson. 2002. Sustainability and panarchies. Pp. 63–102 in *Panarchy: Understanding transformations in human and natural systems*, edited by L. H. Gunderson and C. S. Holling. Washington, D.C.: Island Press.

Jansson, B. O., and H. Velner. 1995. The Baltic: The sea of surprises. Pp. 292–374 in *Barriers and bridges to the renewal of ecosystems and institutions*, edited by L. H. Gunderson, C. S. Holling, and S. S. Light. New York: Columbia University Press.

Kauffman, S. 1995. *At home in the universe: The search for the laws of self-organization and complexity*. New York: Oxford University Press.

Lee, K. N. 1993. *Compass and gyroscope*. Washington, D.C.: Island Press.

Light, S. S., L. H. Gunderson, and C. S. Holling. The Everglades: Evolution of management in a turbulent ecosystem. Pp. 103–168 in *Barriers and bridges to the renewal of ecosystems and institutions*, edited by L. H. Gunderson, C. S. Holling, S. S. Light. New York: Columbia University Press.

Ludwig, D., R. Hilborn, and C. Walters. 1993. Uncertainty, resource exploitation, and conservation: Lessons from history. *Science* 260:17, 36.

Ludwig, D., D. D. Jones, and C. S. Holling. 1978. Qualitative analysis of insect outbreak systems: Spruce budworm and forest. *Journal of Animal Ecology* 47:315–332.

Ludwig, J., D. Tongway, D. Freudenberger, and J. Noble. 1997. *Landscape ecology: Function and management principles from Australia's rangelands*. CSIRO: Collingwood, Australia.

Ostrom, E. 1990. *Governing the commons*. New York: Cambridge University Press.

————. 1995. Designing complexity to govern complexity. In *Property rights and the environment*, edited by S. Hanna and M. Munasinghe. Washington, D.C.: Beijer International Institute and World Bank.

Peterson G. D., C. R. Allen, and C. S. Holling. 1998. Ecological resilience, biodiversity, and scale. *Ecosystems* 1:6–18.

Scheffer, M. 1998. *Ecology of shallow lakes*. London: Chapman and Hall.

Scheffer, M., S. H. Hosper, M.-L. Meijer, B. Moss, and E. Jeppesen. 1993. Alternative equilibria in shallow lakes. *Trends in Ecology and Evolution* 8:275–279.

Sinclair, A. R. E., P. D. Olsen, and T. D. Redhead. 1990. Can predators regulate small mammal populations? Evidence from house mouse outbreaks in Australia. *Oikos* 59:382–392.

Sousa, W. P., and J. H. Connell. 1983. On the evidence needed to judge ecological stability or persistence. *American Naturalist* 121:789–825.

Tenner, E. 1996. *Why things bite back: Technology and the revenge of unintended consequences*. New York: Knopf.

Tilman, D., and J. A. Downing. 1994. Biodiversity and stability in grasslands. *Nature* 367:363–365.

Tilman, D., D. Wedin, and J. Knops. 1996. Productivity and sustainability influenced by biodiversity in grassland ecosystems. *Nature* 379:718–720.

Walker, B. H. 1992. Biological diversity and ecological redundancy. *Conservation Biology* 6:18–23.

————. 1995. Conserving biological diversity through ecosystem resilience. *Conservation Biology* 9:747–752.

Walker, B. H., A. Kinzig, and J. Langridge. 1999. Plant attribute diversity, resilience and ecosystem function: The nature and significance of dominant and minor species. *Ecosystems* 2:95–113.

Walker, B. H., J. L. Langridge, and F. McFarlane. 1997. Resilience of an Australian savanna grassland to selective and nonselective perturbations. *Australian Journal of Ecology* 22:125–135.

Walker, B. H., D. Ludwig, C. S. Holling, and R. M. Peterman. 1981. Stability of semi-arid savanna grazing systems. *Journal of Ecology* 69:473–498.

Walters, C. J. 1986. *Adaptive management of renewable resources*. New York: McGraw Hill.

Weaver, J., P. C. Paquet, and L. Ruggiero. 1996. Resilience and conservation of large carnivores in the Rocky Mountains. *Conservation Biology* 10:964–976.

Young, M., and B. J. McCay. 1995. Building equity, stewardship and resilience into market-based property rights systems. In *Property rights and the environment*, edited by S. Hanna and M. Munasinghe. Washington, D.C.: Beijer International Institute and World Bank.

Young, M. D. 1992. *Sustainable investment and resource use*. Paris: Parthenon Publishing Group.

List of Contributors

Stephen R. Carpenter
Center for Limnology
University of Wisconsin
680 N. Park Street
Madison, WI 53706, USA

Kathryn L. Cottingham
Department of Biological Sciences
Dartmouth College
6044 Gilman, Room 214
Hanover, NH 03755-3576, USA

Terry J. Done
Australian Institute of Marine Science
PMB 3, Townsville MC
Townsville, 4810, Queensland, Australia

Carl Folke
Department of Systems Ecology
Stockholm University
S-106 91 Stockholm, Sweden

Lance H. Gunderson
Department of Environmental Studies
Emory University
Atlanta, GA 30322 USA

C. S. (Buzz) Holling
16871 Sturgis Circle
Cedar Key, FL 32625, USA

AnnMari Jansson
Department of Systems Ecology
Stockholm University
S-106 91 Stockholm, Sweden

Bengt-Owe Jansson
Department of Systems Ecology
Stockholm University
S-106 91 Stockholm, Sweden

Don Ludwig
Department of Mathematics
University of British Columbia
Vancouver, B.C., Canada V6T 1Z2

Ariel E. Lugo
International Institute of Tropical Forestry
USDA Forest Service
P.O. Box 25000
Río Piedras, Puerto Rico 00928-5000

Tim R. McClanahan
The Wildlife Conservation Society
Coral Reef Conservation Project
P.O. Box 99470
Mombasa, Kenya

Sandra Molina Colón
Pontifical Catholic University of Puerto Rico
2250 Las Americas Suite 570
Ponce, Puerto Rico 00731-6382

Peter G. Murphy
Department of Botany and Plant Pathology
Michigan State University
East Lansing, Michigan 48824

Garry D. Peterson
Center for Limnology
University of Wisconsin
680 N. Park Street
Madison, WI 53706, USA

Nicholas C. V. Polunin
Department of Marine Sciences and Coastal Management
The University
Newcastle upon Tyne, NE1 7RU, UK

Lowell Pritchard Jr.
Department of Environmental Studies
Emory University
Atlanta, GA 30322 USA

Frederick N. Scatena
International Institute of Tropical Forestry
USDA Forest Service
P.O. Box 25000
Río Piedras, Puerto Rico 00928-5000

Whendee Silver
Department of Environmental Science, Policy, and Management
University of California
Berkeley, CA 94720-3110 USA

Brian H. Walker
Sustainable Ecosystems
Commonwealth Scientific and Industrial Research Organization
GPO Box 284
Canberra, ACT 2601, Australia

Carl J. Walters
Department of Zoology
University of British Columbia
Vancouver, B.C., Canada V6T 1W5

Scope Series List

SCOPE 1–59 are now out of print. Selected titles from this series can be down-loaded free of charge from the SCOPE Web site (http://www.icsu-scope.org).

SCOPE 1:   *Global Environment Monitoring,* 1971, 68 pp
SCOPE 2:   *Man-made Lakes as Modified Ecosystems,* 1972, 76 pp
SCOPE 3:   *Global Environmental Monitoring Systems (GEMS): Action Plan for Phase I,* 1973, 132 pp
SCOPE 4:   *Environmental Sciences in Developing Countries,* 1974, 72 pp
SCOPE 5:   *Environmental Impact Assessment: Principles and Procedures,* Second Edition, 1979, 208 pp
SCOPE 6:   *Environmental Pollutants: Selected Analytical Methods,* 1975, 277 pp
SCOPE 7:   *Nitrogen, Phosphorus and Sulphur: Global Cycles,* 1975, 129 pp
SCOPE 8:   *Risk Assessment of Environmental Hazard,* 1978, 132 pp
SCOPE 9:   *Simulation Modelling of Environmental Problems,* 1978, 128 pp
SCOPE 10: *Environmental Issues,* 1977, 242 pp
SCOPE 11: *Shelter Provision in Developing Countries,* 1978, 112 pp
SCOPE 12: *Principles of Ecotoxicology,* 1978, 372 pp
SCOPE 13: *The Global Carbon Cycle,* 1979, 491 pp
SCOPE 14: *Saharan Dust: Mobilization, Transport, Deposition,* 1979, 320 pp
SCOPE 15: *Environmental Risk Assessment,* 1980, 176 pp
SCOPE 16: *Carbon Cycle Modelling,* 1981, 404 pp
SCOPE 17: *Some Perspectives of the Major Biogeochemical Cycles,* 1981, 175 pp
SCOPE 18: *The Role of Fire in Northern Circumpolar Ecosystems,* 1983, 344 pp
SCOPE 19: *The Global Biogeochemical Sulphur Cycle,* 1983, 495 pp
SCOPE 20: *Methods for Assessing the Effects of Chemicals on Reproductive Functions, SGOMSEC 1,* 1983, 568 pp
SCOPE 21: *The Major Biogeochemical Cycles and their Interactions,* 1983, 554 pp
SCOPE 22: *Effects of Pollutants at the Ecosystem Level,* 1984, 460 pp

SCOPE 47: *Long-Term Ecological Research. An International Perspective*, 1991, 312 pp

SCOPE 48: *Sulphur Cycling on the Continents: Wetlands, Terrestrial Ecosystems and Associated Water Bodies*, 1992, 345 pp

SCOPE 49: *Methods to Assess Adverse Effects of Pesticides on Non-target Organisms, SGOMSEC 7*, 1992, 264 pp

SCOPE 50: *Radioecology after Chernobyl*, 1993, 367 pp

SCOPE 51: *Biogeochemistry of Small Catchments: a Tool for Environmental Research*, 1993, 432 pp

SCOPE 52: *Methods to Assess DNA Damage and Repair: Interspecies Comparisons, SGOMSEC 8*, 1994, 257 pp

SCOPE 53: *Methods to Assess the Effects of Chemicals on Ecosystems, SGOMSEC 10*, 1995, 440 pp

SCOPE 54: *Phosphorus in the Global Environment: Transfers, Cycles and Management*, 1995, 480 pp

SCOPE 55: *Functional Roles of Biodiversity: a Global Perspective*, 1996, 496 pp

SCOPE 56: *Global Change, Effects on Coniferous Forests and Grasslands*, 1996, 480 pp

SCOPE 57: *Particle Flux in the Ocean*, 1996, 396 pp

SCOPE 58: *Sustainability Indicators: a Report on the Project on Indicators of Sustainable Development*, 1997, 440 pp

SCOPE 59: *Nuclear Test Explosions: Environmental and Human Impacts*, 1999, 304 pp

SCOPE Executive Committee 2001–2004

President:
Dr. Jerry M. Melillo (USA)

1st Vice-President:
Prof. Rusong Wang (China-CAST)

2nd Vice-President:
Prof. Bernard Goldstein (USA)

Treasurer:
Prof. Ian Douglas (USA)

Secretary-General:
Prof. Osvaldo Sala

Members:
Prof. Himansu Baijnath (South Africa-IUBS)
Prof. Holm Tiessen
Prof. Manuwadi Hungspreugs
Prof. Venugopalan Ittekkot
Prof. Reynaldo Victoria

Index

Island Press Board of Directors

Chair
HENRY REATH
President, Collectors Reprints, Inc.

Vice-Chair
VICTOR M. SHER
Miller Sher & Sawyer

Secretary
DANE A. NICHOLS
Chair, The Natural Step

Treasurer
DRUMMOND PIKE
President, The Tides Foundation

ROBERT E. BAENSCH
Director, Center for Publishing,
New York University

MABEL H. CABOT
President, MHC & Associates

DAVID C. COLE
Owner, Sunnyside Farms

CATHERINE M. CONOVER

CAROLYN PEACHEY
Campbell, Peachey & Associates

WILL ROGERS
President, Trust for Public Land

CHARLES C. SAVITT
President, Center for Resource
Economics/Island Press

SUSAN E. SECHLER
Senior Advisor on Biotechnology
Policy, The Rockefeller Foundation

PETER R. STEIN
Managing Partner,
The Lyme Timber Company

RICHARD TRUDELL
Executive Director, American Indian
Resources Institute

DIANA WALL
Director and Professor,
Natural Resource Ecology Laboratory,
Colorado State University

WREN WIRTH
President, The Winslow Foundation